Solid Dielectric Horn Antennas

For a complete listing of the *Artech House Antenna Library*,
turn to the back of this book.

Solid Dielectric Horn Antennas

Carlos Salema
Carlos Fernandes
Rama Kant Jha

Artech House
Boston • London

Library of Congress Cataloging-in-Publication Data
Salema, Carlos.
Solid dielectric horn antennas / Carlos Salema, Carlos Fernandes, Rama Kant Jha.
p. cm. — (Artech House antenna library)
Includes bibliographical references and index.
ISBN 0-89006-911-5 (alk. paper)
1. Antennas, Horn. 2. Dielectric devices. I. Fernandes, Carlos.
II. Jha, Rama Kant. III. Title. IV. Series.
TK7871.6.S25 1998
621.382'4—dc21 98-10761
CIP

British Library Cataloguing in Publication Data
Salema, Carlos
Solid dielectric horn antennas
1. Dielectric wave guides 2. Antennas, Horn 3. Dielectric wave guides—Design 4. Antennas, Horn—Design
I. Title II. Fernandes, Carlos III. Jha, Rama Kant
621.3'81331
ISBN 0890069115

Cover design by Jennifer L. Stuart

685 Canton Street
Norwood, MA 02062

International Standard Book Number: 0-89006-911-5
Library of Congress Catalog Card Number: 98-10761

10 9 8 7 6 5 4 3 2 1

To our families

Contents

Foreword

Dielectric waveguides are among the earliest structures investigated for the guidance of electromagnetic waves. Their use as antennas emerged during World War II, then in the 1960s, and again more recently a number of new types of dielectric radiating structure have been produced following the invention of the dielguide antenna which was first described in 1966.

Surprisingly, there have been very few books devoted solely to the subject of dielectric antennas and the book by Professor Salema and his colleagues is timely. It is ideally suited to the research worker who seeks a rigorous account of the theoretical principles underlying the design of a dielectric antenna and the inclusion of experimental results, many of which have been obtained by the authors, lends confidence to approximations that are usually required in order to predict the radiation pattern. The chapters are supported by a sound bibliography, and the inclusion of a chapter setting out the fundamentals of electromagnetic waves makes the book ideal as a postgraduate text.

Although the authors have had microwave and millimeter wave applications primarily in mind when preparing their text, it is important to remember that many of the principles also apply to optical fiber waveguides, therefore the book is also likely to be found on the shelves of research workers in that field. The authors are to be congratulated in producing what is a definitive text that is likely to remain an authority for many years ahead.

Peter Clarricoats
Queen Mary and Westfield College
University of London

Acknowledgments

The authors wish to record their sincere thanks and appreciation to their colleagues, research scholars, and supporting staff both at Instituto de Telecomunicações and Instituto Superior Técnico, Lisboa, Portugal and at the Electronics Engineering Department, Banaras Hindu University, Varanesi, India, who helped in different ways in completing this daunting task. At the risk of omitting many names they would like to explicitly mention the special help of Professors Fernando Pereira and Afonso Barbosa, Mr. Manuel Menezes de Sequeira, Mr. Paulo Francês, Mr. David Lemaire, Mr. Jorge Silva and Mr. Vasco Fred in Lisboa, and Drs. S. P. Singh, B. Jha, T. Tiwari, S. K. Pathak, and Sri Ravishankar in Varanesi.

We also acknowledge Junta Nacional de Investigação Cientıifica e Tecnológica, Lisboa, Portugal, for the financial support to C. Fernandes (partly through Project PBIC/ C/ TIT/ 2501/ 95) and to R. K. Jha (PRAXIS XXI visiting scientist fellowship), and Instituto de Telecomunicações and Instituto Superior Téecnico for providing a very congenial and stimulating environment.

Last but not the least, we would like to thank the silent partners — our families — for their immense forbearance.

Carlos Salema
Carlos Fernandes
Rama Kant Jha

Acknowledgments

Chapter 1

Introduction

1.1 DIELECTRIC ANTENNAS

In spite of the spread in the use of optical fibers, radio communications still play a significant role in telecommunications. The current trend in mobile communications ensures that this role is likely to remain important and may even grow in the foreseeable future.

Radio spectrum congestion and the need for an ever-increasing bandwidth place stiffer requirements on antennas. Demand for mobile communications on one hand, and cellular broadcasting, on the other hand, have pushed radio communications first into microwave and later into millimeter waves. At these frequencies resistive losses introduce a penalty in metallic antennas which may be compounded by increasing manufacturing costs. The availability of high performance low-loss dielectric materials makes dielectric antennas a viable candidate for millimeter wave applications.

Although there is plenty of literature on dielectric antennas there are few books exclusively dedicated to the subject. In most cases the treatment of dielectric antennas tends to be confined to one or two relatively brief chapters which almost invariably restrict the subject to dielectric rods and lenses. Even if these antennas have been used in the past and continue to be used nowadays very successfully there are other relatively unexplored shapes that may conceivably be used as antennas.

In this book we approach the basic radiation mechanisms involved in some lesser known dielectric structures; in this way we hope to provide insight that should enable a preliminary antenna design. This could be a first step, which may considerably reduce the turnaround time, in a design procedure that ultimately would require the use of standard computer-aided design (CAD) packages [1, 2, 3]. Alternatively it could provide information for the development of improved CAD packages.

Useful dielectric antennas may be based on simple geometries: cylindrical, conical, parallelepipedic, pyramidal, spherical, and hemispherical. More elaborate shapes also exist, often as variations of the previous geometries and mostly with axis-symmetry: lenses are one important subgroup of the dielectric antennas with more elaborate shapes.

Figure 1.1 shows some basic dielectric antenna configurations that will be addressed in this book.

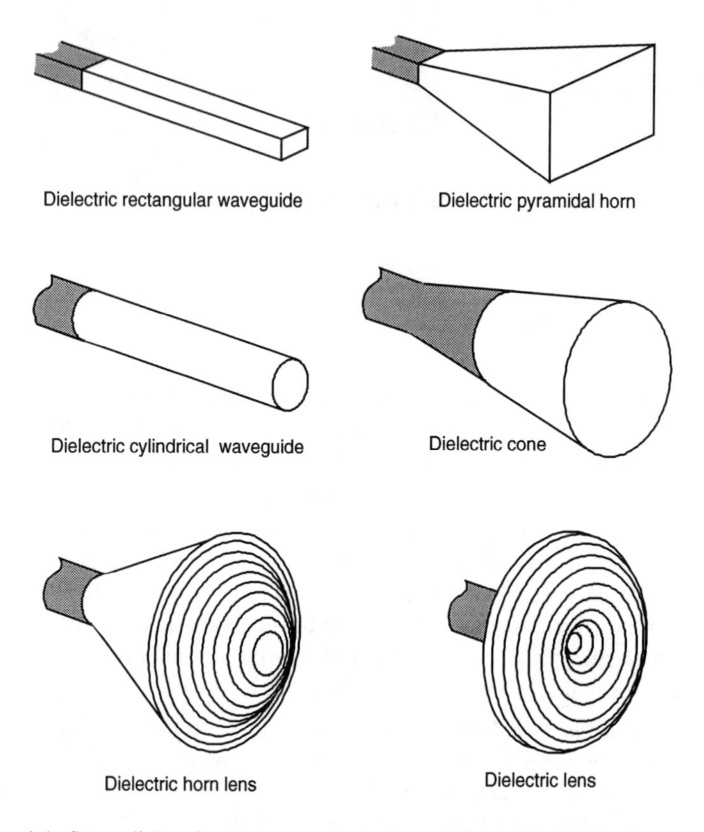

Figure 1.1: Some dielectric antenna configurations addressed in this book.

Dielectric antennas usually include two parts: the dielectric structure and the feed, also known as the launcher. The dielectric structure may be solid, hollow, or may have a metal core. In most applications the dielectric structure is homogeneous, but it may be nonhomogeneous with step or with continuous refractive index variation. The dielectric structure may be embedded in the feed structure or may be mounted on top of the feed aperture; it may even be located at some distance from the feed.

Depending on the objectives, the dielectric structure may be fed by an open-ended metallic waveguide, by a metallic horn, by wire or printed dipoles or loops, or by slots.

Metal shields that do not significantly affect the fields in the dielectric structure may be added to improve the radiation pattern reducing its sidelobe level.

At least for some feed/dielectric antenna combinations, the dielectric structure may be viewed as a load that modifies the original metallic feed radiation characteristics, to obtain some improvements, such as an increased directivity and/or the polarization purity, a reduction of the sidelobe level, or a modified main lobe shape or scanning capability, just to mention a few. Such antennas are sometimes referred to as dielectric loaded antennas [4, 5].

We will make the following distinction between dielectric antennas and dielectric loaded antennas: in the former metal plays no essential part in the shaping of the radiating fields, even though it may be indispensable as a launcher or as a shield; whereas in the latter metal interacts closely with the dielectric to produce the required field structure.

With a few exceptions, dielectric antennas are most useful at frequencies well above 1 GHz. They are usually lightweight, low cost, easily machinable and have good sealing and corrosion properties. Further, we may expect that future plastic dielectrics will be not only of high quality but will also be qualified to function in space; plastic dielectrics will enable these antennas to be of use at microwave, millimeter wave, and even higher frequency ranges.

Some common applications of dielectric antennas are:

- Feeds for metallic reflectors and lenses;
- Loads to modify the radiation pattern of metal horns;
- Broadband antenna replacement of metal horns;
- Phased array elements;
- Antenna element in an integrated circuits receiver [6].

Many other uses of these antennas are given in [4].

According to the geometry, the aperture fields of dielectric antennas can be represented as an appropriate expansion in terms of plane, cylindrical, or spherical waves that are reflected and transmitted at the boundaries between the different dielectric regions. In the case of lenses, where the dielectric boundary may have a complex shape and the aperture dimensions are of the order of 15 to 20 wavelengths, a reasonable approximation of the aperture fields may be obtained using geometrical optics. Once the fields are known, the corresponding far-field radiation pattern can be calculated using Kirchhoff-Huygens formulas.

1.2 BRIEF HISTORY OF DIELECTRIC ANTENNAS

An exhaustive review of dielectric and dielectric loaded antennas was carried out by Chatterjee [4]. Recently dielectric loaded horns and dielectric feeds were also extensively reviewed by Olver [5]. Earlier reviews of some of these antennas are due to James [7] and Love [8].

For the purpose of this section, limited to references deemed relevant for the content of the book, we will divide dielectric antennas in the following three broad categories: (1) rods, (2) horns, and (3) lenses.

1.2.1 Dielectric Rods

A dielectric rod antenna is made up of a length of dielectric rod, with circular or rectangular cross section, fed at one end by a launcher, usually a metal waveguide. To improve the radiation pattern and the impedance matching the rod is often tapered down toward the free end. Further modifications of the basic structure include shaping of the rod open end into some special form [9, 10] and the introduction of a cylindric metallic core [11].

The earliest work appears to be due to Hondros and Debye [12], who carried out a theoretical investigation of microwave propagation along a solid dielectric cylinder as early as 1910. Work on these antennas remained neglected or in dormant state until World War II, basically because of the lack of high frequency technology.

Some frequently referred early workers are Mueller [13] and Watson and Horton [14, 15].

Assuming that the power flow is mostly confined to the dielectric, Marcatili [16] and Goell [17] derived the fields for an infinite dielectric prism with rectangular cross section. Their formulation, in some cases complemented with the introduction of the concept of "effective dielectric constant" [18], is still used as the basis for rectangular rod analysis [19, 20].

With the exception of Wilkes [21], who approached the dielectric rod as a lens, theoretical analyses were based on the following principles:

1. Scalar Huygens formula on the side walls [22];
2. Equivalence theorem, applied to the whole rod [23, 24];
3. Vector Kirchhoff-Huygens formula, taking two apertures, hence the name "two aperture theory" [7, 25, 26, 27]: the free end where the fields are taken as the rod modal fields and the launcher end assumed to radiate unaffected by the presence of the rod, except for power being converted into rod modes.

While some authors claimed that the whole surface was responsible for the radiation, others contended that the radiation came only from the two ends, the cylinder serving only to convey surface waves from one end to the other. Zucker [26], Collin [27], and Andersen [28] helped to clear up the matters by pointing out that for a finite rod the whole surface is responsible for the radiation pattern and that simplified assumptions on the distribution of surface currents were responsible for discrepancies in earlier results. In fact, because of the reflections and diffractions at the launcher and the free end surface, currents are not smooth as previously assumed.

The excitation efficiency of dielectric rods is dealt with in [29, 30, 31] among others for the circular section and in [19, 20] for the rectangular section.

1.2.2 Dielectric Horns

A dielectric horn may be conceived as a dielectric rod whose cross section gradually expands toward the free end. The dielectric horn may be solid or hollow. The cross section may be circular or rectangular. In the latter case the expansion may be either in the direction of one side (sectoral horn) or of both sides (pyramidal horn). In most cases the expansion of the cross section is linear but occasionally it may also be exponential.

Dielectric horns are usually excited by metallic waveguides or horns. In this case the antenna will be classified as a dielectric horn when the dielectric structure plays the major role in setting up the fields that give rise to the radiation pattern. Shielded dielectric horns, where a metal shield restricts the radiating aperture mostly to the dielectric free end, without otherwise affecting the field structure in the dielectric, will be classified as dielectric horns.

Dielectrics may also be used as lining, coating, or filling of metal horns in order to improve the resulting radiation pattern or the impedance matching. Many other variations may be found in the literature. In all these cases, however, metal plays the dominant role and hence these antennas will be named as dielectric loaded metal horns and will not be considered further in this book.

The low permittivity dielectric cone was originally employed, in conjunction with a metal horn, as an improved feed for a reflector antenna [32]. Later the solid dielectric cone came on its own as an antenna [33, 34, 35, 36, 37, 38].

A theoretical analysis of the solid sectoral dielectric horn, both in the E-plane and the H-plane forms, is described in [39, 40, 41, 42, 43]. The radiation characteristics of a modified rhombic dielectric plate antenna, which may be described as a solid dielectric H-plane sectoral horn modified by the addition of a simple triangular shaped lens, are given in [44], without theoretical details.

Work on the hollow dielectric sectoral horn is reported in Jha and Aas [45, 46].

The H-plane multilayer dielectric horn was dealt with in [47]. The objective was to find the appropriate relative dielectric constants, axial lengths, and the flare angles for the horn so that the waves can travel through the dielectric and reach the front aperture with only minimum phase differences achieving maximum aperture efficiency.

The first analysis of the solid pyramidal dielectric horn, of low permittivity, dates back to 1974 [48].

A comparative study of hollow dielectric and metallic pyramidal horns with different flare angles and axial lengths [49] leads to the conclusion that dielectric horns if properly designed may be more directive than their metal counterparts. Another work [50] reports that the E-plane patterns of the hollow pyramidal dielectric horns have lower sidelobes than the H-plane patterns and also are at lower power levels.

The radiation patterns of the dielectric diagonal horn, a special case of the pyramidal horn, was recently reported [51].

1.2.3 Dielectric Lenses

Microwave use of dielectric lenses seems to date back to the early days of the electromagnetic waves experiments, associated with the verification of the optical properties of the electromagnetic waves at 60 GHz [52]. However, it was not until World War II that lenses gained interest as antenna elements.

Lenses have been widely used to convert spherical phase fronts into planar phase fronts across an aperture to enhance its directivity much like parabolic reflectors. For this purpose lenses present an advantage over reflectors in that the feed is located behind the lens, thus eliminating aperture blockage by the feed and supporting struts, with no need for offset solutions.

The main disadvantage of dielectric lenses is the bulk at microwave frequencies; zoning has been used to reduce lens thickness and further to reduce lens aberrations in scanning and multibeam applications, but surface stepping reduces the lens inherently large bandwidth. Metal plate lenses have been used as a less massive alternative, with the added advantage of allowing extra degrees of freedom for the design.

Nowadays there is a renewed interest on dielectric lenses due to the rapidly growing number of new applications at millimeter waves, where lens physical dimensions have acceptable sizes. Besides, very low loss dielectrics are available, and present-day numerically controlled machines enable low-cost development of sophisticated lenses.

In one of the earliest dielectric lens antenna applications, a homogeneous lens was designed to produce a wide-angle scanning lobe [53]. Scanning and multibeam applications motivated a number of subsequent works on dielectric lenses [54, 55, 56], some of them based on nonhomogeneous lens configurations [57].

Homogeneous lenses also have been used as phase front correctors for horns. In most cases [58, 59, 60, 61, 63] the lens is mounted as a cap on a hollow metallic horn. In this configuration the two lens surfaces can be used to design for two simultaneous conditions.

Besides phase correction, lenses may be designed to further control the taper of the field distribution at the lens aperture [61, 64, 65] or to shape the amplitude of the output beam in special applications [66].

The aperture of a solid dielectric horn can be shaped into a lens to modify or improve some radiation characteristics [62]. For instance, the aperture efficiency of a solid dielectric horn may be improved by correcting the aperture phase error. Alternatively we may use a lens to shape the amplitude of the output beam or to improve the cross-polarization performance. Because there is only one lens surface to shape in a solid dielectric horn, only one of the previous design targets can be considered at a time.

In most cases lens dimensions are much larger than the wavelength, and design is based on the geometrical optics formulation, resulting in a reasonably frequency independent behavior. Snell's refraction law and a path length condition (or eventually an energy conservation condition) are then used to define the lens surfaces in the limit as the wavelength tends to zero. Depending on the lens shape, diffraction effects, which are not accounted for in the geometrical optics formulation, may give rise to discrepancies in the final pattern. The actual far-field diffraction pattern is computed in a second step by aperture integration of the lens aperture fields, obtained by ray tracing.

Risser [54] gives the basic mathematical formulation for lens design, but the subject has been expanded and updated in subsequent antenna textbooks [24, 56, 67, 68, 69]. Sletten [70] presents software to design the most important types of lenses along with the respective feeds.

In all the cases referred to so far the design refers to axis-symmetric lenses. Exceptions are Westcott [65], who generalized a previously developed formulation for two-reflector antenna design for arbitrary shaped dielectric lenses, and Fernandes [66] who described a three-dimensional (3D) amplitude shaping dielectric lens for emerging mobile communications applications at millimeter waves.

1.3 GENERAL APPROACH TO RADIATION PATTERN ANALYSIS

In general the analytical formulation and solution of the excitation of a finite length dielectric structure by an open-ended metallic waveguide or horn is not trivial. For a constant cross-section structure the fields at the launcher end may be obtained as a superposition of a finite number of discrete modes plus a continuous spectrum wave with decaying amplitude away from the feed open end [71, 72, 73]. When these waves reach the dielectric open end, they excite reflected propagating waves, as well as another continuous-spectrum wave. The amplitudes of all these waves depend on the geometry and on the electric parameters, and can be obtained only by solving the full wave problem.

There is no available formulation that yields an exact analytical solution for the full wave problem. The Wiener-Hopf formulation [74] can provide an exact solution for a certain class of problems involving junctions of semi-infinite structures. In [75] it was applied to the excitation of a grounded dielectric slab by a terminated parallel plate metallic waveguide. It was also applied to a cylindrical dielectric rod fed by a circular metallic waveguide carrying the circular symmetric TM_{01} mode [76]. The generalization for the case of asymmetric excitation, which is of interest for this book, is overwhelming. Truncation of a Wiener-Hopf structure to a finite length yields in general a Wiener-Hopf equation of the modified type [77, 78], which only in restricted cases is amenable to an approximate solution. Direct application of the Wiener-Hopf formulation for the rectangular cross–section dielectric or for a flared structure excited by a constant cross–section waveguide is not possible.

Amenable analytical solutions require some simplifications to the full problem. One possibility could be to replace the waveguide aperture by a sheath of equivalent currents [79] and use it to obtain the discrete and the continuous-spectrum waves excited in the dielectric structure when assumed infinite. Integration of the resulting near fields over a finite portion of the infinite structure using the Kirchhoff-Huygens formulas (see Chapter 2) may yield a satisfactory approximation of the far-field pattern of the real antenna.

Other simplified approaches may be found in the literature. One, due to James [7], assumes that the launcher radiation pattern is unperturbed by the presence of the dielectric except for a reduction in power amplitude due to power being converted in the dielectric structure modes. As detailed in Chapter 3, the results provided by this method may be significantly improved if the launcher phase pattern is suitably modified to take into account the presence of the dielectric structure.

Another approach, put forward by Clarricoats and Salema [34] for very low permittivity ($\bar{\varepsilon} \leq 1.2$) dielectrics and described in detail in Section 5.6, assumes that the radiation pattern of the launcher is null for angles from boresight to a given trapping angle (related to the permittivity) and essentially the unperturbed launcher pattern beyond this angle.

These two methods converge when the excitation efficiency is reasonably high (say above about 80%) and the directivity of the dielectric structure is much higher and its sidelobe level much lower than that of the launcher. In this case the interference between the launcher and the dielectric structure main lobe can safely be neglected, whereas at the launcher's sidelobe level the opposite is true.

From the above references a simplified approach to the analysis of the radiation pattern of most dielectric antennas, and in effect of many other aperture antennas, may be summarized as follows:

1. The antenna is divided into two main components: the dielectric structure and the launcher;
2. The fields of the dielectric structure, assumed infinite in extent and isolated in free space, are derived;
3. The fields in the actual finite dielectric structure are taken as equal to the previously derived ones or to the sum of a forward wave plus a backward wave caused by reflections where the infinite structure was truncated to produce the real antenna;
4. The radiation pattern due to the dielectric structure is computed from the field distributions using either the equivalence theorem or some form of the Kirchhoff–Huygens formulas;
5. The contribution from the launcher is added to the radiation pattern of the dielectric structure.

In shielded dielectric antennas radiation is mainly due to the front aperture, a feature that greatly simplifies the application of the Kirchhoff-Huygens formulas. Although the launcher does not contribute directly to the radiation it nevertheless determines which dielectric modes are excited and may reach the aperture. For shielded dielectric horns the design of the throat region is critical for single-mode operation required for high-performance antennas.

1.4 SCOPE OF THE BOOK

The increased use of the higher microwave and millimeter wave frequencies offers a new opportunity for dielectric antennas that may well replace classic metal antennas at least in some cases.

The book, which is heavily based on the authors', and their groups', research and development (R&D) work, emphasizes the basic radiation mechanisms of a few basic types of dielectric antennas aiming at deriving their radiation patterns from known parameters: dimensions and permittivity. The stress is on the analysis rather than on the design, although this topic is also addressed to a limited extent.

Aimed primarily at researchers and postgraduate students in the field of microwave and millimeter wave antennas, the book may provide valuable information for some applications even if it is not intended as a dielectric antenna design manual.

In some chapters the text is heavily loaded with mathematics. Although the authors are aware that this may scare quite a few readers, they felt that omitting such information would make it very difficult for the willing reader to pursue further this work. This is all the more important since some dielectric antennas still have many little-known aspects that deserve further research work both at the theoretical and at the experimental level.

Most examples referred to are of prototype antennas operating in the X band. This is due to the authors' workshop and measurement facilities, particularly for phase measurements, and by no means implies that these prototypes configure practical proposals for real-life antennas at this wavelength. In fact dielectric antennas, although usable and useful at X band, really come into their own at millimeter wave frequencies when their size and cost compares favorably with alternatives providing similar performance.

The book is organized in seven chapters in addition to three appendixes.

The introduction (Chapter 1) is followed by a discussion of the fundamentals of electromagnetic waves in bounded and unbounded media (Chapter 2) and a review of radiation patterns for rectangular and circular apertures with some simple field distributions. The solid rectangular dielectric rod and the principles behind the shielded dielectric antennas are dealt with in Chapter 3. Chapter 4 treats solid sectoral and pyramidal dielectric horns as gradually expanding rectangular waveguides and provides a simple design method for shielded dielectric antennas. A discussion of the dielectric cylinder and the dielectric cone makes up Chapters 5 and 6.

In all these chapters the subject is approached in a similar way. The field equations are solved in the dielectric structure and aperture fields are derived. For a number of launchers the excitation efficiency is evaluated using modal matching techniques. The radiation pattern of the dielectric antenna is derived by integration of the Kirchhoff-Huygens formulas, and the estimated launcher effects are added to produce the final radiation pattern.

Chapter 7 is devoted to single-surface dielectric lenses, used to modify the radiation pattern of dielectric horns or waveguides, either by acting on the aperture phase distribution or by shaping the output beam. The design procedure involves two steps: in the first one geometric optics is used to calculate the lens profile according to design specifications. In the second step, physical optics are used to compute the radiation pattern and account for the diffraction effects.

REFERENCES

[1] Hewlett–Packard EEsof, *HFSS 5.0*, Computer Program.

[2] Kolundzija, B. (1997), *Proceedings of the IEEE APS International Symposium*, July, Vol. 3, pp. 1822–1825.

[3] Toland, B., Liu, C., and Ingerson, P., (1997), Design and Analysis of Arbitrary Shaped Dielectric Antennas, *Microwave Journal*, Vol. 40, No. 4, May, pp. 278–286.

[4] Chatterjee, R. (1985), *Dielectric and Dielectric Loaded Antennas*, Research Studies Press Ltd., Letchworth, Hertfordshire, England.

[5] Olver, A. D., Clarricoats, P. J. B., Kishk, A. A., and Shafai, L., (1994), *Microwave Horns and Feeds*, IEEE Press, New York.

[6] Yao, C., Schwartz, S. E., and Blumenstock, B. J., (1982), Monolithic Integration of a Dielectric mmw Antenna and Mixer Diode: An Embryonic mmw Ic, *IEEE Transactions on Microwave Theory and Techniques*, Vol. MTT-30, No. 8, pp. 1241–1246.

[7] James, J. R., (1967), Theoretical Investigation of Cylindrical Dielectric Rod Antennas, *Proceedings of the IEE*, Vol. 144, pp. 309–319.

[8] Love, A. W., (1976) , *Reflector Antennas*, IEEE Press, New York.

[9] Vedavaty, T. S., (1972), *Dielectric Coated Spherically Tipped Metal Cone Aerials Excited in Unsymmetric Hybrid Mode*, Ph.D. Thesis, Indian Institute of Science, Bangalore, India.

[10] Chatterjee, R., and Vedavaty, T. S., (1975), Dielectric Coated Spherically Tipped Conducting Cone Aerials Excited in Unsymmetric Hybrid Mode at Microwave Frequencies, *Journal of the Indian Institute of Science*, Vol. 57, No. 11, pp. 399–425.

[11] Hersch, W., (1960), Surface Wave Aerial, *IEE Monograph No. 363E*, pp. 202–212.

[12] Hondros, D., and Debye, P., (1910), Electromagnetische Wellen an Dielektrischen Drahten, *Ann. der Physik*, Vol. 32, No. 8, pp. 465–476.

[13] Mueller, G. E., and Tyrrel, W. A., (1947), Polyrod Antennas, *Bell System Technical Journal*, Vol. 26, No. 4, pp. 837–851.

[14] Watson, R. B., and Horton, C. W., (1948), The Radiation Pattern of Dielectric Rods – Experiment and Theory, *Journal of Applied Physics*, Vol. 19, pp. 661–670.

[15] Watson, R. B., and Horton, C. W., (1948), On the Calculation of Radiation Patterns of Dielectric Rods, *Journal of Applied Physics*, Vol. 19, pp. 836–837.

[16] Marcatili, E. A. J., (1969), Dielectric Rectangular Wave Guide and Directional Coupler for Integrated Optics, *Bell System Technical Journal*, Vol. 48, pp. 2071–2102.

[17] Goell, J. E., (1969), A Circular Harmonic Computer Analysis of Rectangular Dielectric Waveguides, *Bell System Technical Journal*, Vol. 48, No. 7, pp. 2133–2160.

[18] McLevige, W., Itoh, T., and Mittra R., (1975), New Waveguide Structures for Millimeter-Wave and Optical Integrated Circuits, *IEEE Transactions on Microwave Theory and Techniques*, Vol. MTT-23, No.10, pp. 788–794.

[19] Sen, T. K., and Chatterjee, R., (1978), Rectangular Dielectric Rod at Microwave Frequencies- Pt-I, Theoretical and Experimental Determination of Launching Efficiency, *Journal of the Indian Institute of Science*, Vol. 60, No. 5, pp. 193–210.

[20] Sen, T. K., and Chatterjee, R., (1978), Rectangular Dielectric Rod at Microwave Frequencies– Part II, Radiation Characteristics, *Journal of the Indian Institute of Science*, Vol. 60(5), pp. 211–225.

[21] Wilkes, G., (1948), Wavelength Lenses, *Proceedings of the IRE* ,Vol. 36, pp. 206–212.

[22] Halliday, D. F., and Kiely, D. G., (1947), Dielectric Rod Aerials, *Journal of the IEE*, Vol. 94, Pt. IIIA, pp. 610–618.

[23] Chatterjee, R., and Chatterjee, S. K., (1957), Some Investigations on Dielectric Rod Aerials, Part II, *Journal of Indian Institute of Science*, Vol. 39(2), pp. 134–140.

[24] Fradin, A. Z., (1961), *Microwave Antennas*, Pergamon Press, London.

[25] Brown, J., and Spector, J. O., (1957), The Radiating Properties of Endfire Aerials, *Proceedings of the IEE*, Vol. 104(B), pp. 27–34.

[26] Zucker, F. J., (1969), Surface- and Leaky-Wave Antennas, in *Antenna Engineering Handbook*, ed. Jasik, McGraw-Hill, New York.

[27] Collin, R. E., and Zucker, F. J., (1969), *Antenna Theory*, Vol. 2, McGraw-Hill, New York.

[28] Andersen, J. B., (1970), *Metallic and Dielectric Antennas*, Polyteknisk Forlag, Denmark.

[29] Duncan, J. W., (1959), The Efficiency of Excitation of a Surface Wave on a Dielectric Cylinder, *IRE Transactions on Microwave Theory and Techniques*, Vol. MTT-7, pp. 257–265.

[30] Wenger, N. C., (1965), The launching of surface waves on a cylindrical reactive surface, *IEEE Transactions on Antennas and Propagation*, Vol. AP-13, pp. 125–134.

[31] Brown, J., and Stechera, H. S., (1962), Launching of an Axial Cylindrical Surface Wave, *Proceedings of the IEE*, Vol. 109(S), pp. 18–25.

[32] Bartlett, H. E., and Moseley, R. E., (1966), Dielguides — Highly Efficient Low Noise Antenna Feeds, *Microwave Journal*, Vol. 9, pp. 53–58.

[33] Clarricoats, P. J. B. and Salema, C. E. R. C., (1971), Propagation and Radiation Characteristics of Low Permittivity Dielectric Cones, *Electronics Letters*, Vol. 7, No. 17, pp. 483–485.

[34] Clarricoats, P. J. B., and Salema, C. E. R. C., (1972), Influence of the Launching Horn on the Radiation Characteristics of a Dielectric Cone Feed, *Electronics Letters*, Vol. 8, No. 8, pp. 200–202.

[35] Salema, C. E. R. C., and Clarricoats, P. J. B., (1972), Radiation Characteristics of Dielectric Cones, *Electronics Letters*, Vol. 8, No. 16., pp. 414–416

[36] Salema, C. E. R. C., (1972), *Theory and Design of Dielectric Cone Antennas*, Ph. D. Thesis, University of London.

[37] Clarricoats, P. J. B., and Salema, C. E. R. C., (1973) Antennas Employing Conical Dielectric Horns, Part I, Proceedings of the IEE, Vol. 120, pp. 741–749.

[38] Clarricoats, P. J. B., and Salema, C. E. R. C, (1973), Antennas Employing Conical Dielectric Horns Part II, *Proceedings of the IEE*, Vol. 120, pp. 750–756.

[39] Jha, B., and Jha, R. K., (1984), Surface Fields of H-Plane Sectoral Dielectric Horn Antenna, *Proceedings of the 15th European Microwave Conference*.

[40] Jha, B., (1984), *Sectoral Dielectric Horn Antennas*, Ph.D. Thesis, Banaras Hindu University, Varanasi, India.

[41] Singh, A. K., Jha, B., and Jha, R.K., (1990), Near-Field Analysis of H-Plane Hollow Sectoral Dielectric Horn Antennas, *International Journal of Electronics*, Vol. 68, No. 6, pp. 1055–1061.

[42] Singh, A. K., Jha, B., and Jha, R. K., (1991), Near-Field Analysis of E-Plane Sectoral Solid Dielectric Horn Antennas, *International Journal of Electronics*, Vol. 71, No. 4, pp. 697–706.

[43] Singh, A. K., (1991), *Study of Some Aspects of Sectoral Dielectric Horn Antennas*, Ph. D. Thesis, Banaras Hindu University, Varanasi, India.

[44] Ohtera, I., and Ujiie, H., (1981), Radiation Perfomances of a Modified Rhombic Dielectric Plate Antenna, *IEEE Transactions on Antennas and Propagation*, Vol. AP-29, No.4, pp. 660-662.

[45] Jha, R. K., and Aas, J. A., (1977), H-plane Sectoral Dielectric Horn Antenna, *Proceedings of the IEE-IERE (India)*, Vol. 15, No. 2, pp. 69–77.

[46] Jha, R. K., Jha, L., and Misra, D. K., (1976), Study of Gain in Dielectric Horns, *Journal of Electronics Engineering (India)*, Vol. VII, pp. 203–205.

[47] Bayat, A., (1993), *Studies of Some Aspects of H-Plane Sectoral Multilayer Solid Dielectric Horn Antennas*, Ph.D. Thesis, Banaras Hindu University, Varanasi, India.

[48] Brooking, N., Clarricoats, P. J. B., and Olver, A. D., (1974) Radiation Pattern of Pyramidal Dielectric Waveguides, *Electronics Letters*, Vol. 10, No. 3, pp. 33–34.

[49] Jha, R. K., Misra, D.K. and Jha, L., (1975), Comparative Study of Dielectric and Metallic Horn Antennas, *IEEE – Conference, India Section, Annual Convention Record, pp. 29.*

[50] Misra, D. K., Jha, L., and Jha, R. K., (1975), Experimental Investigation of Pyramidal Dielectric Horn Antennas, *IEEE — Antennas and Propagation International Symposium Digest*, Illinois, U.S.A., pp. 16–18

[51] Singh, S. P., Tiwari, T., Tiwari, V. M., and Jha, R. K., (1994), Radiation Patters of Metal and Dielectric Walls Diagonal Horn Antennas, *International Journal of Electronics*, No. 6, Vol. 76, pp. 1195–1203.

[52] Sengupta, D. L., and Sarkar, T. K., (1995), Microwave and Millimeter Wave Research Before 1900 and the Centenary of the Horn Antenna, *Proceedings of the 25th European Microwave Conference*, Bologna, Italy, pp. 903–909.

[53] Friedlander, F. G., (1946), A Dielectric Lens Aerial for Wide Angle Beam Scanning, *Journal of the IEE*, Vol. 93(3A), pp. 658–662.

[54] Silver, S. (ed.), (1949), *Microwave Antenna Theory And Design*, McGraw Hill, New York.

[55] Shinn, D. H., (1955), The Design of Zoned Dielectric Lens for Wide-angle Scanning, *Marconi Review*, No. 117, pp.37–47.

[56] Lo, Y. T., and Lee, S. W. (Eds.), (1988), *Antenna Handbook Theory Applications and Design*, Van Nostrand, New York.

[57] Sanford, J. R., (1994), Scattering by Spherically Stratified Microwave Lanses, *IEEE Transactions on Antennas and Propagation*, Vol. AP-42, No. 5, pp. 690–698.

[58] Kildal, P. S., (1984), Meniscus–Lens–Corrected Corrugated Horn: A Compact Feed for a Cassegrain Antenna, *Proceedings of the IEE*, Vol. 131(H), pp. 390–394.

[59] Clarricoats, P. J. B., and Saha, P. K., (1969), The Radiation Pattern of Lens Corrected Conical Scalar Horn, *Electronics Letters*, Vol. 5, pp. 592–593.

[60] Olver, A. D., and Saleeb, A. A., (1983), Improved Radiation Characteristics of Conical Horns with Plastic-foam Lenses, *Proceedings of the IEE*, Vol. 130(H), pp. 197–202.

[61] Kildal, P. S., and Jacobson, K., (1984) Scalar Horn with Shaped Lens Improves Cassegrain Efficiency, *IEEE Transactions on Antennas and Propagation*, Vol. AP-32, No. 10, pp. 1094–1100.

[62] Olver, A. D., and Philips, B., (1993), Integrated Lens with Dielectric Horn Antenna, *Electronics Letters*, Vol. 29, No. 13, pp. 1150–1152.

[63] Khoury, K., and Heane, G. W., (1991), High Performance Lens Horn Antennae for the Millimetre Bands, *IEE Colloquium*, Digest No. 1991/012, pp. 10/1–10/9.

[64] Lee, J. J., (1983), Dielectric Lens Shaping and Coma- Correction Zoning Pt-I: Analysis, *IEEE Transactions on Antennas and Propagation*, Vol. AP-31, No. 1, pp. 211–216.

[65] Westcott, B. S., (1986), General Dielectric–Lens Shaping Using Complex Coordinates, *Proceedings of the IEE*, Vol. 133(H), pp. 122–126.

[66] Fernandes, C. A., Francês, P. O. and Barbosa, A. M., (1994), Shaped Coverage of Elongated Cells at mm Waves Using a Dielectric Lens Antenna, *Proceedings of the 25th European Microwave Conference*, pp. 66–70.

[67] Kraus, J. D., (1950), *Antennas*, McGraw-Hill, New York.

[68] Jasik, H., (1961), *Antenna Engineering Hand Book*, McGraw-Hill, New York.

[69] Cornbleet, S., (1994), *Microwave And Geometrical Optics*, Academic Press, London.

[70] Sletten, C. J., (1988), *Reflectors And Lens Antennas — Analysis And Design Using Personal Computers*, Artech House, Norwood, pp. 215–260.

[71] Collin, R. E., (1960), *Field Theory of Guided Waves*, McGraw-Hill, New York.

[72] Tamir, T., and Oliner, A., (1963), Guided Complex Waves (Part I and Part II) *Proc. IEE*, Vol. 110, No. 2, pp. 310–334.

[73] Shevchenko, V., (1971), *Continuous Transitions in Open Waveguides*, Golem Press, Boulder, Colorado.

[74] Weinstein, L. A., (1969), *The Theory of Diffraction and the Factorization Method*, Golem Press, Boulder, Colorado.

[75] Angulo, C. M., and Chang, W., (1959), The Launching of Surface Waves by a Parallel Plate Waveguide, *IRE Transactions on Antennas and Propagation*, Vol. AP-7, pp. 359–368.

[76] Angulo, C. M., and Chang, W., (1958), The Excitation of a Dielectric Rod by a Cylindrical Waveguide, *IRE Transactions on Microwave Theory and Techniques*, Vol. MTT-6, pp. 389–393.

[77] Noble, B., (1958), *Methods Based on the Wiener-Hopf Technique*, Pergamon Press, London.

[78] Mittra, R., and Lee, S. W., (1971), *Analytical techniques in the Theory of Guided Waves*, McGraw-Hill, New York.

[79] Fernandes, C., and Barbosa A., (1990), Complex Wave Radiation from a Sheath Helix Excited by the Circular Waveguide TE_{11} Mode, *Journal of Electromagnetic Waves and Applications*, Vol. 4, No. 6, pp. 549–571.

Chapter 2

Fundamentals of Electromagnetic Waves

2.1 INTRODUCTION

Although this book is intended for readers with a knowledge of basic electromagnetics, we thought it would be useful, for consistency in notation and easy referencing, to dedicate a chapter to the fundamentals of propagation of electromagnetic waves in bounded and unbounded media and radiation from apertures. Most of the material presented here may be found in classical textbooks such as Harrington [1] and Silver [2].

Starting with a review of Maxwell's equations, in both differential and integral forms, we consider the field behavior at boundaries, the power flow, and we derive the field from field potentials. The general equations are later specialized for time harmonic fields and the wave equations are derived. Reflection and refraction of uniform plane waves are treated next and the concepts of wave impedance, reflection, and transmission coefficients are introduced.

Following the proof of the equivalence theorem, the vector Kirchhoff-Huygens formulas are presented and their use in computing the radiation pattern of aperture antennas is discussed. Special cases for plane and spherical cap apertures are introduced and practical forms of the Kirchhoff-Huygens formulas are derived.

Rectangular and circular apertures with uniform and cosine illumination, both with and without quadratic phase errors, are used as examples to review basic design principles and techniques that in later chapters are applied to dielectric horns.

The chapter ends with a few concepts commonly used in antenna work and with a very brief discussion of materials referred to in this book.

2.2 MAXWELL'S EQUATIONS

Maxwell's equations, sometimes called electromagnetic equations, express the fundamental relations between electromagnetic fields and their sources: currents and charges. In the MKSA [1] rationalized system of units that is used throughout this book, they can be written as

$$\nabla \times \mathcal{E} = -\frac{\partial \mathcal{B}}{\partial t} \tag{2.1}$$

$$\nabla \times \mathcal{H} = \mathcal{J} + \frac{\partial \mathcal{D}}{\partial t} \tag{2.2}$$

$$\nabla \cdot \mathcal{D} = \rho \tag{2.3}$$

$$\nabla \cdot \mathcal{B} = 0 \tag{2.4}$$

where

- $\mathcal{D}$ is the total electric displacement density or total electric flux density (coulomb per square meter);
- $\mathcal{E}$ is the electric field (volt per meter);
- $\mathcal{B}$ is the magnetic flux density (weber per square meter);
- $\mathcal{H}$ is the magnetic field (ampere per meter);
- $\mathcal{J}$ is the electric current density (ampere per square meter);
- ρ is the electric charge density (coulomb per cubic meter).

All the above quantities refer to instantaneous values.

Applying Gauss's and Stokes's theorems to (2.1) and (2.2) and to (2.3) and (2.4), respectively, we obtain Maxwell's equations in integral form:

$$\oint_C \mathcal{E} \cdot d\boldsymbol{\ell} = -\frac{\partial}{\partial t}\left(\int_S \mathcal{B} \cdot \boldsymbol{n}\, dS\right) \tag{2.5}$$

$$\oint_C \mathcal{H} \cdot d\boldsymbol{\ell} = \int_S \left(\mathcal{J} + \frac{\partial \mathcal{D}}{\partial t}\right) \cdot \boldsymbol{n}\, dS \tag{2.6}$$

$$\int_S \mathcal{D} \cdot \boldsymbol{n}\, dS = \int_V \rho\, dV \tag{2.7}$$

$$\int_S \mathcal{B} \cdot \boldsymbol{n}\, dS = 0 \tag{2.8}$$

Following Gauss's and Stokes's theorems, in the previous equations V is a bounded volume, enclosed by surface S, and $\boldsymbol{n}$ is the unit vector outward normal to S.

Electric charges and currents are related through the continuity equation:

$$\nabla \cdot \mathcal{J} = -\frac{\partial \rho}{\partial t} \tag{2.9}$$

[1] The MKSA system is based on four fundamental units: *meter*, *kilogram*, *second*, and *ampere*

In a linear, homogeneous, and isotropic medium we have the following relations:

$$\mathcal{D} = \varepsilon \mathcal{E} \quad (2.10)$$

$$\mathcal{B} = \mu \mathcal{H} \quad (2.11)$$

$$\mathcal{J} = \sigma \mathcal{E} \quad (2.12)$$

where ε (farad per meter) is the medium permittivity or dielectric constant, μ (henry per meter) is the medium permeability or magnetic constant, and σ (ohm $^{-1}$ per meter) is the medium conductivity.

In vacuum we have

- $\varepsilon = \varepsilon_0 = \frac{1}{36\pi} \times 10^{-9}$ (farad per meter)
- $\mu = \mu_0 = 4\pi \times 10^{-7}$ (henry per meter)
- $\sigma = \sigma_0 = 0$

Perfect conductors have $\sigma = \infty$ and perfect dielectrics $\sigma = 0$.

Permittivity and permeability of the linear materials (sometimes also called the simple materials) are often expressed as relative permittivity (or relative dielectric constant) $\bar{\varepsilon}$ and relative permeability (or relative magnetic constant) $\bar{\mu}$:

$$\bar{\varepsilon} = \frac{\varepsilon}{\varepsilon_0} \quad (2.13)$$

$$\bar{\mu} = \frac{\mu}{\mu_0} \quad (2.14)$$

For all nonmagnetic materials $\bar{\mu} = 1$.

2.3 BOUNDARY CONDITIONS

Consider the boundary surface $S_{1,2}$ between media 1 and 2. Denote their respective permittivities, permeabilities, and conductivities by ε_1 and ε_2, μ_1 and μ_2, and σ_1 and σ_2, respectively. Assume a cylinder with volume V and surface S embracing both media: the cylinder side walls are normal to the boundary surface (Figure 2.1) and have an area much smaller than that of the bases.

Applying Maxwell's integral equation (2.7) to the cylinder we get

$$\int_S \mathcal{D} \cdot \boldsymbol{n}\, dS = \int_V \rho\, dV \quad (2.15)$$

If we now shrink the cylinder dimensions in such a way that the base is reduced to an area dS we get

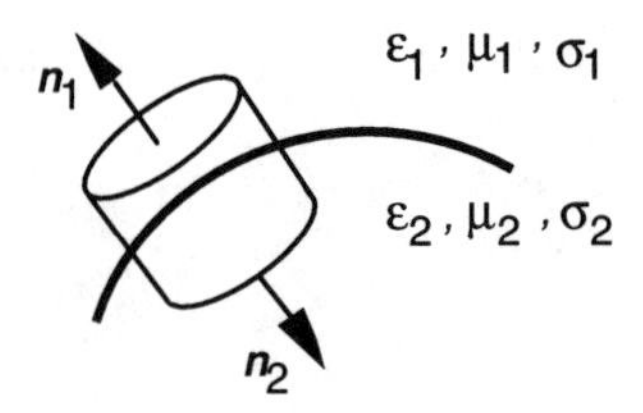

Figure 2.1: Boundary conditions: elementary cylinder for the application of Gauss's theorem.

$$(\boldsymbol{\mathcal{D}}_2 \cdot \boldsymbol{n}_2 + \boldsymbol{\mathcal{D}}_1 \cdot \boldsymbol{n}_1)\, dS = \rho\, dV \tag{2.16}$$

or, noting that, $\boldsymbol{n}_2 = \boldsymbol{n}$ and $\boldsymbol{n}_1 = -\boldsymbol{n}$:

$$(\boldsymbol{\mathcal{D}}_2 - \boldsymbol{\mathcal{D}}_1) \cdot \boldsymbol{n} = \chi_e \tag{2.17}$$

where χ_e is the electric surface charge density.

Equation (2.17) may be expressed as follows: the normal component of $\boldsymbol{\mathcal{D}}$ is discontinuous across a boundary by an amount equal to the electric surface charge density.

The same reasoning applied to Maxwell's integral equation (2.8) leads us to the conclusion that the normal component of $\boldsymbol{\mathcal{B}}$ is continuous across the boundary.

Consider again the boundary surface $S_{1,2}$ between two media with parameters ε_1, μ_1, and σ_1, and ε_2, μ_2, and σ_2, respectively. This time assume a contour C across this boundary and a positive sense around it such as represented in Figure 2.2.

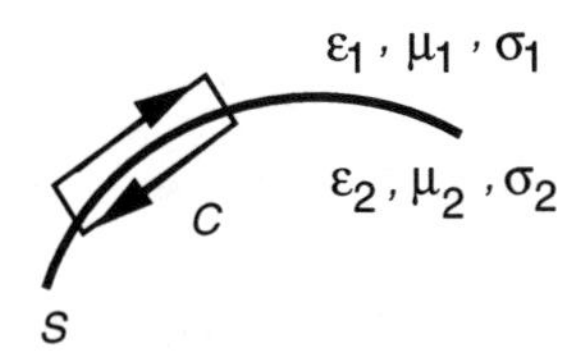

Figure 2.2: Boundary conditions: contour for the application of Stokes's theorem.

Applying Maxwell's integral equation (2.6) to the contour in Figure 2.2 we get

$$\oint_C \boldsymbol{\mathcal{H}} \cdot d\boldsymbol{\ell} = \int_S \left(\boldsymbol{\mathcal{J}} + \frac{\partial \boldsymbol{\mathcal{D}}}{\partial t} \right) \cdot \boldsymbol{n}\, dS \tag{2.18}$$

Shrinking the contour dimensions in such a way that the length tangential to S is reduced to $d\ell$ and the length normal to S is always much smaller than the tangential one we get in the usual cases

$$(\mathcal{H}_1 - \mathcal{H}_2) \cdot d\boldsymbol{\ell} = 0 \tag{2.19}$$

which means that the tangential component of the magnetic field is continuous across the boundary.

There is a notable exception to the above: when one of the media is a perfect conductor. In this case although the integration surface tends to zero the integral tends to a finite value — the surface current per unit width.

Similarly, Maxwell's integral equation (2.5) leads to the continuity of the tangential component of $\mathcal{E}$.

2.4 POYNTING THEOREM

The energy carried by electromagnetic waves may be obtained from Maxwell's equations. Scalar multiplication of (2.1) by $\mathcal{H}$ and of (2.2) by $\mathcal{E}$ and subtraction yields

$$\mathcal{H} \cdot (\nabla \times \mathcal{E}) - \mathcal{E} \cdot (\nabla \times \mathcal{H}) = -\mathcal{H} \cdot \frac{\partial \mathcal{B}}{\partial t} - \mathcal{E} \cdot \mathcal{J} - \mathcal{E} \cdot \frac{\partial \mathcal{D}}{\partial t} \tag{2.20}$$

which for a linear, homogeneous, and isotropic media can be rewritten as

$$\nabla \cdot (\mathcal{E} \times \mathcal{H}) = -\frac{\partial}{\partial t}\left(\frac{\epsilon E^2}{2} + \frac{\mu H^2}{2}\right) - \mathcal{E} \cdot \mathcal{J} \tag{2.21}$$

where E and H denote, respectively, the modulus of vectors $\mathcal{E}$ and $\mathcal{H}$.

The cross product of vectors $\mathcal{E}$ and $\mathcal{H}$ is known as the Poynting vector and is usually denoted by $\mathcal{S}$:

$$\mathcal{S} = \mathcal{E} \times \mathcal{H} \tag{2.22}$$

Applying Gauss's theorem on a volume V bounded by surface S to (2.21) we get

$$\int_S (\mathcal{E} \times \mathcal{H}) \cdot \boldsymbol{n}\, dS = -\int_V \mathcal{E} \cdot \mathcal{J}\, dV - \frac{\partial}{\partial t}\int_V \left(\frac{\varepsilon E^2}{2} + \frac{\mu H^2}{2}\right) dV \tag{2.23}$$

Equations (2.21) and (2.23) express the Poynting theorem in its differential and integral forms, respectively. This theorem may be stated as follows: the flux of the normal component of the Poynting vector $\mathcal{S}$ across a surface S that encloses a volume V is equal to the ohmic power losses plus the variation of the electric and magnetic energy in V. The minus sign on the right-hand side of (2.23) coupled with the normal on S pointing outward from V indicates that — for ohmic losses to take place in volume V and for electric and magnetic energy in the same volume to increase — the flux of the Poynting vector must be negative. Bearing in mind the law of conservation of energy applied to V, it follows that the Poynting vector represents the power per unit area.

2.5 VECTOR POTENTIALS

To derive the electromagnetic fields from the field sources — charges and currents — one usually makes use of two auxiliary functions, $\mathcal{A}$ and ϕ, known as the magnetic vector and electric scalar potentials, respectively. Alternatively, use could be made of another pair of functions, $\mathcal{F}$ and ϕ_m, known as the electric vector and the magnetic scalar potentials, respectively.

Let us recall Maxwell's equations (2.1) through (2.4) in their differential form for a linear, homogeneous, and isotropic media. Since $\nabla \cdot \mathcal{B} = 0$ we can state that $\mathcal{B}$ is the curl of a new vector $\mathcal{A}$, the magnetic vector potential:

$$\mathcal{B} = \nabla \times \mathcal{A} \tag{2.24}$$

Introducing (2.24) in (2.1) and noting that $\mathcal{B} = \mu\mathcal{H}$ we get

$$\nabla \times (\mathcal{E} + \frac{\partial \mathcal{A}}{\partial t}) = 0 \tag{2.25}$$

Since a vector whose curl is zero can be derived from the gradient of a scalar function we will introduce the electric scalar potential ϕ such that

$$\mathcal{E} + \frac{\partial \mathcal{A}}{\partial t} = -\nabla\phi \tag{2.26}$$

The choice of the minus sign corresponds to the usual convention for the electric field, for which the positive direction goes from increasing to decreasing values of potential.

From (2.26) we can write

$$\mathcal{E} = -\nabla\phi - \frac{\partial \mathcal{A}}{\partial t} \tag{2.27}$$

Equations (2.24) and (2.27) show how fields $\mathcal{E}$ and $\mathcal{B}$ are derived from vector potential $\mathcal{A}$ and scalar potential ϕ. The remaining electromagnetic fields are derived from these using (2.10) and (2.11).

We will now look for expressions relating the vector and the scalar potentials with the field sources, that is, currents and charges. Introducing (2.24) and (2.27) in (2.2) and using (2.10) we get for a lossless medium

$$\nabla \times (\nabla \times \mathcal{A}) = \mu\mathcal{J} - \varepsilon\mu\frac{\partial \nabla\phi}{\partial t} - \varepsilon\mu\frac{\partial^2 \mathcal{A}}{\partial t^2} \tag{2.28}$$

Recalling (A.11) and noting that $c^2 = 1/(\omega\varepsilon)$, as will be proved later, (2.28) becomes

$$\nabla^2\mathcal{A} - \frac{1}{c^2}\frac{\partial^2 \mathcal{A}}{\partial t^2} = -\mu\mathcal{J} + \nabla\left(\nabla \cdot \mathcal{A} + \frac{1}{c^2}\frac{\partial \phi}{\partial t}\right) \tag{2.29}$$

On the other hand, introducing (2.27) in (2.1) we get

$$\nabla^2 \phi + \frac{\partial}{\partial t}(\nabla \cdot \mathcal{A}) = -\frac{\rho}{\varepsilon} \tag{2.30}$$

Let us go back to (2.27) and (2.24). It can be easily seen that the fields remain unchanged when the potentials are modified as follows:

$$\phi \to \phi_{\text{mod}} = \phi + \frac{\partial \psi}{\partial t} \tag{2.31}$$

$$\mathcal{A} \to \mathcal{A}_{\text{mod}} = \mathcal{A} - \nabla \psi \tag{2.32}$$

where ψ is any well-behaved function of x, y, z and t.

If we impose the condition that ψ has to obey the wave equation

$$\nabla^2 \psi - \frac{1}{c^2}\frac{\partial^2 \psi}{\partial t^2} = 0 \tag{2.33}$$

from (2.32) and (2.31) we get

$$\nabla \cdot \mathcal{A}_{\text{mod}} = \nabla \cdot \mathcal{A} - \nabla^2 \psi \tag{2.34}$$

$$= \nabla \cdot \mathcal{A} - \frac{1}{c^2}\frac{\partial^2 \psi}{\partial t^2} \tag{2.35}$$

$$\frac{\partial \phi_{\text{mod}}}{\partial t} = \frac{\partial \phi}{\partial t} + \frac{\partial^2 \psi}{\partial t^2} \tag{2.36}$$

Introducing the value of $\frac{\partial^2 \psi}{\partial t^2}$ obtained from (2.36) into (2.35) we get

$$I = \nabla \cdot \mathcal{A}_{\text{mod}} + \frac{1}{c^2}\frac{\partial \phi_{\text{mod}}}{\partial t} = \nabla \cdot \mathcal{A} + \frac{1}{c^2}\frac{\partial \phi}{\partial t} \tag{2.37}$$

which means that I is invariant relative to the transformations defined by (2.32) and (2.31) [3]. We can, quite arbitrarily, set $I = 0$ and obtain the Lorentz condition:

$$\nabla \cdot \mathcal{A} + \frac{1}{c^2}\frac{\partial \phi}{\partial t} = 0 \tag{2.38}$$

Imposing the Lorentz condition on (2.29) and (2.30) we finally get

$$\nabla^2 \mathcal{A} - \frac{1}{c^2}\frac{\partial^2 \mathcal{A}}{\partial t^2} = -\mu \mathcal{J} \tag{2.39}$$

$$\nabla^2 \phi - \frac{1}{c^2}\frac{\partial^2 \phi}{\partial t^2} = -\frac{\rho}{\varepsilon} \tag{2.40}$$

The Lorentz condition sets the vector potential $\mathcal{A}$ apart from the scalar potential ϕ. While the first depends only on the currents, the latter depends solely on the charges. In addition, using the Lorentz condition we may express the electric field only in terms of the vector potential:

$$\mathcal{E} = c^2 \int_0^t \nabla(\nabla \cdot \mathcal{A})dt - \frac{\partial \mathcal{A}}{\partial t} \tag{2.41}$$

Equations (2.41) and (2.24) show another consequence of the Lorentz condition: the fields can now be deduced simply from the knowledge of the impressed currents, regardless of the charges.

Making use of the symmetric form of Maxwell's equation we can define another set of vector and scalar potentials, the electric vector potential $\mathcal{F}$ and the magnetic scalar potential ϕ_m, from which to derive the electromagnetic fields. Applying symmetry to (2.24) and (2.27) it follows that

$$\varepsilon\mathcal{E} = -\nabla \times \mathcal{F} \tag{2.42}$$

$$\mathcal{H} = -\nabla\phi_m - \frac{\partial \mathcal{F}}{\partial t} \tag{2.43}$$

Similarly the Lorentz condition (2.38) is now

$$\nabla \cdot \mathcal{F} + \frac{1}{c^2}\frac{\partial \phi_m}{\partial t} = 0 \tag{2.44}$$

and from (2.39) and (2.40) the equations relating the electric vector and the scalar magnetic potentials to the magnetic current and charge densities $\mathcal{J}_m$ ρ_m are

$$\nabla^2\mathcal{F} - \frac{1}{c^2}\frac{\partial^2 \mathcal{F}}{\partial t^2} = -\varepsilon\mathcal{J}_m \tag{2.45}$$

$$\nabla^2\phi_m - \frac{1}{c^2}\frac{\partial^2 \phi_m}{\partial t^2} = -\frac{\rho_m}{\mu} \tag{2.46}$$

As above the Lorentz condition enables us to write the magnetic field simply in terms of the electric vector potential:

$$\mathcal{H} = c^2 \int_0^t \nabla(\nabla \cdot \mathcal{F})dt - \frac{\partial \mathcal{F}}{\partial t} \tag{2.47}$$

Integration of (2.39) and (2.40) leads to the delayed potentials:

$$\mathcal{A}(P, t) = \frac{\mu}{4\pi}\int_{V'} \frac{\mathcal{J}\left(P', t - \frac{r}{c}\right)}{r} dV' \tag{2.48}$$

$$\phi(P, t) = \frac{1}{4\pi\varepsilon}\int_{V'} \frac{\rho\left(P', t - \frac{r}{c}\right)}{r} dV' \tag{2.49}$$

where V' is the domain of existence of $\mathcal{J}$, and r is the distance between P and P'.

2.6 TIME HARMONIC FIELDS

For time harmonic fields, that is, for time variation of the type $\exp(j\omega t)$, where $\omega = 2\pi f$ is the angular frequency, for an homogeneous, linear, and isotropic medium, Maxwell's equations in symmetric form can be rewritten, using complex notation, as

$$\nabla \times \boldsymbol{E} = -\boldsymbol{J}_m - j\omega\mu \boldsymbol{H} \tag{2.50}$$

$$\nabla \times \boldsymbol{H} = \boldsymbol{J}_e + (\sigma + j\omega\varepsilon)\boldsymbol{E} \tag{2.51}$$

$$\nabla \cdot \boldsymbol{E} = \frac{\rho}{\varepsilon} \tag{2.52}$$

$$\nabla \cdot \boldsymbol{H} = \frac{\rho_m}{\mu} \tag{2.53}$$

where $\boldsymbol{J}_e$ and $\boldsymbol{J}_m$ are the impressed electric and magnetic current densities. The reader will notice that here we have distinguished in the electric current density the conduction current density $\sigma \boldsymbol{E}$ and the impressed current density $\boldsymbol{J}_e$ and that there is no magnetic equivalent of the conduction current density.

The main advantage of using the symmetric form of Maxwell's equations is the possibility to substitute in any derived equation $\boldsymbol{E}$ for $\boldsymbol{H}$, $\boldsymbol{H}$ for $-\boldsymbol{E}$, $(\sigma + j\omega\varepsilon)$ for $j\omega\mu$ and vice versa, $\boldsymbol{J}_e$ for $\boldsymbol{J}_m$ and $\boldsymbol{J}_m$ for $-\boldsymbol{J}_e$, ρ/ε for ρ_m/μ, and ρ_m/μ for $-\rho/\varepsilon$.

In the following, when assuming for a variable a time variation of the type $\exp(j\omega t)$, the symbol for the variable always denotes the root mean square (rms) amplitude, as in Harrington [1]. Had we chosen the peak value, as many authors prefer, then all time average power (energy) values would have to be divided by 2.

Under the above conditions, the laws of continuity of charges become

$$\nabla \cdot (\sigma \boldsymbol{E} + \boldsymbol{J}_e) = -j\omega\rho \tag{2.54}$$

$$\nabla \cdot \boldsymbol{J}_m = -j\omega\rho_m \tag{2.55}$$

Applying Stokes's and Gauss's theorems to (2.50) and (2.51) and to (2.52) and (2.53), respectively, and dropping impressed currents we obtain Maxwell's equations in integral form:

$$\oint_C \boldsymbol{E} \cdot d\boldsymbol{\ell} = -j\omega\mu \int_S \boldsymbol{H} \cdot \boldsymbol{n}\, dS \tag{2.56}$$

$$\oint_C \boldsymbol{H} \cdot d\boldsymbol{\ell} = (\sigma + j\omega\varepsilon) \int_S \boldsymbol{E} \cdot \boldsymbol{n}\, dS \tag{2.57}$$

$$\int_S \boldsymbol{E} \cdot \boldsymbol{n}\, dS = \int_V \frac{\rho}{\varepsilon}\, dV \tag{2.58}$$

$$\int_S \boldsymbol{H} \cdot \boldsymbol{n}\, dS = 0 \tag{2.59}$$

where, as noted above, V is a bounded volume, enclosed by surface S, and $\boldsymbol{n}$ is the unit vector outward normal to S.

The law of continuity of electric charges may also be expressed in the integral form. Applying Gauss's theorem to (2.54) we get

$$\int_S (\sigma \boldsymbol{E} + \boldsymbol{J}_e) \cdot \boldsymbol{n} \, dS = -j\omega \int_V \rho \, dV \tag{2.60}$$

Expressed in terms of the vector potentials, the fields become

$$\boldsymbol{E} = \frac{1}{j\omega\varepsilon\mu} \nabla(\nabla \cdot \boldsymbol{A}) - j\omega \boldsymbol{A} \tag{2.61}$$

$$\boldsymbol{H} = \frac{1}{\mu} \nabla \times \boldsymbol{A} \tag{2.62}$$

$$\boldsymbol{E} = -\frac{1}{\varepsilon} \nabla \times \boldsymbol{F} \tag{2.63}$$

$$\boldsymbol{H} = \frac{1}{j\omega\varepsilon\mu} \nabla(\nabla \cdot \boldsymbol{F}) - j\omega \boldsymbol{F} \tag{2.64}$$

2.7 WAVE EQUATIONS

Let us consider a linear, homogeneous, source-free, and isotropic medium and take for the electromagnetic fields a time variation of the type $\exp(j\omega t)$. Applying the curl operator to Maxwell's equation (2.50) and using (2.51) we have

$$\nabla \times (\nabla \times \boldsymbol{E}) - k^2 \boldsymbol{E} = 0 \tag{2.65}$$

where

$$k = \omega\sqrt{\mu\epsilon} \tag{2.66}$$

is the propagation constant, or wave number.

Recalling vector identity (A.11) and (2.52) we get

$$\nabla^2 \boldsymbol{E} + k^2 \boldsymbol{E} = 0 \tag{2.67}$$

Similarly, we get for the magnetic field

$$\nabla^2 \boldsymbol{H} + k^2 \boldsymbol{H} = 0 \tag{2.68}$$

Equations (2.67) and (2.68) are called the wave equations since, as will be shown below, they express the wave nature of propagation of $\boldsymbol{E}$ and $\boldsymbol{H}$. From those equations it follows for rectangular coordinates that each component of the electric and magnetic field vectors must satisfy the scalar Helmholtz equation:

$$\nabla^2 \psi + k^2 \psi = 0 \tag{2.69}$$

In order to demonstrate the wave behavior of the electromagnetic fields, let us take a perfect dielectric ($\sigma = 0$, $\rho = 0$) and a linearly polarized (along the X axis) electric field, that is, an electric field with a single nonzero component (along the X axis), which varies only with z. Equation (2.67) then yields

$$\nabla^2 E_x + k^2 E_x = 0 \tag{2.70}$$

The solutions of this equation are of the type $\exp(\pm jkz)$, that is

$$E_x = E_0 \exp(\pm jkz) \exp(j\omega t) \tag{2.71}$$

According to (2.71), E_x propagates wavelike — that is, assuming the same values along the Z axis after a certain time lag — in the positive direction if we take the negative sign, or in the negative direction if we take the positive sign. The first solution is known as the forward wave and the second as the backward wave. The full solution is a sum of forward and backward waves, the amplitude of which is determined by the boundary conditions.

The velocity of propagation v of the wave is simply

$$\begin{aligned} v &= \mp \frac{w}{k} \\ &= \mp \frac{1}{\sqrt{\varepsilon\mu}} \end{aligned} \tag{2.72}$$

From (2.50) for a source-free medium we find out that

$$H_x = 0 \tag{2.73}$$

$$H_y = \frac{1}{j\omega\mu}\frac{\partial E_x}{\partial z} \tag{2.74}$$

$$= \pm \frac{k}{\omega\mu} E_x \tag{2.75}$$

$$= \pm \sqrt{\frac{\varepsilon}{\mu}} E_x \tag{2.76}$$

$$H_z = 0 \tag{2.77}$$

Therefore we conclude that the magnetic field is also linearly polarized, along the Y axis, that is, normal to the electric field. The positive sense for the magnetic field is opposite for the forward wave and for the backward wave. Using the Poynting vector it is easy to find out that for the forward wave the power flow is along the positive direction of the Z axis, whereas for the backward wave it is in the negative direction of the same axis.

From (2.71) and (2.74) we conclude that the surfaces on which the phase is constant — the *equiphase surfaces* — are planes (here normal to the Z axis). For this reason the wave is known as a *plane wave*.

When field amplitudes are constant on the equiphase surfaces the wave is known as *uniform*.

When the electric and magnetic fields are both transversal to the direction of propagation the wave is known as *transversal electromagnetic* or TEM. Later in this book we will encounter cases where only the electric field or only the magnetic field is transversal to the direction of propagation: these cases are known as transverse electric (TE) and transverse magnetic (TM) waves, respectively.

Returning to the case of a uniform, plane TEM wave in a perfect dielectric we will define the medium intrinsic impedance Z as

$$Z = \frac{E_x}{H_y} = \sqrt{\frac{\mu}{\varepsilon}} \tag{2.78}$$

If the medium is vacuum, where $\varepsilon_0 = 10^{-9}/36\pi$ (Fm^{-1}) and $\mu_0 = 4\pi \times 10^{-7}$ (Hm^{-1}), the velocity of propagation and the medium impedance become, from (2.72) and (2.78),

$$v_0 = \frac{1}{\sqrt{\varepsilon_0 \mu_0}} \tag{2.79}$$

$$= c \tag{2.80}$$

$$= 2.9979 \times 10^8 \ (\text{ms}^{-1}) \tag{2.81}$$

$$Z_0 = \sqrt{\frac{\mu_0}{\varepsilon_0}} \tag{2.82}$$

$$= 120\pi \ (\text{ohm}) \tag{2.83}$$

where c denotes the velocity of light in vacuum, as usual.

The locus with time of the end point of the instantaneous electric field vector in a fixed plane — normal to the direction of propagation — defines the wave polarization. In the general case, this locus is an ellipse and the polarization is named elliptical. Cases of particular interest in antenna work are the circular and the linear polarizations. It can be easily demonstrated that an elliptically polarized wave can be obtained by adding two linearly polarized waves with appropriate amplitude and phase.

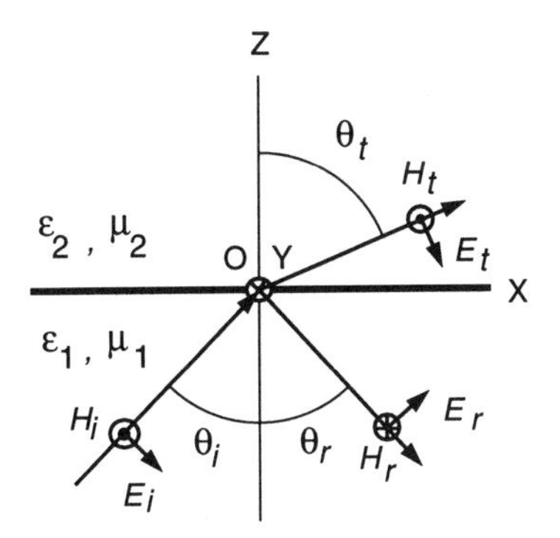

Figure 2.3: Schematic representation of the oblique incidence of a TEM wave on the plane boundary between two media.

2.8 REFLECTION AND REFRACTION OF UNIFORM PLANE WAVES

In free space, at large distance from the source, an electromagnetic wave can locally be assimilated to a uniform, TEM, plane wave, where the electric and magnetic field vectors are normal and mutually normal to the direction of propagation. The stored energies are equally divided between the electric and magnetic fields.

In the following we will consider in some detail the phenomena linked to the incidence of a uniform plane wave on the plane boundary between two media with different characteristics.

Consider a plane boundary between media 1 and 2, lying on XOY (Figure 2.3) and a uniform, TEM, linearly polarized plane wave, the incident wave, coming from medium 1 with a direction of propagation at an angle θ_i with the normal to the boundary. The electric field of the incident wave lies in the plane of incidence (the plane normal to boundary that contains the direction of propagation) and the magnetic field is parallel to the boundary. Such polarization is commonly known as *transverse magnetic*, *parallel*, or *vertical polarization.*

Because of the discontinuity some energy, in the form of a reflected wave, will be reflected back into medium 1 and some energy, in the form of a transmitted wave, will be transmitted to medium 2.

Assuming there are no surface charges or surface currents on the boundary, the fields can be written as

$$E_{x_1} = E_0 \cos(\theta_i)(1 + \Gamma_{\parallel}) \exp[-jk_1(x \sin\theta_i + z\cos\theta_i)] \tag{2.84}$$

$$E_{x_2} = E_0 \cos(\theta_t) T_{\parallel} \exp[-jk_2(x \sin\theta_t + z\cos\theta_t)] \tag{2.85}$$

$$E_{y_1} = 0 \tag{2.86}$$

$$
\begin{aligned}
E_{y_2} &= 0 && (2.87)\\
E_{z_1} &= E_0 \sin(\theta_i)(1+\Gamma_\|)\exp[-jk_1(x\sin\theta_i + z\cos\theta_i)] && (2.88)\\
E_{z_2} &= E_0 \sin(\theta_t) T_\| \exp[-jk_2(x\sin\theta_t + z\cos\theta_t)] && (2.89)\\
H_{x_1} &= 0 && (2.90)\\
H_{x_2} &= 0 && (2.91)\\
H_{y_1} &= \frac{E_0}{Z_1}(1-\Gamma_\|)\exp[-jk_1(x\sin\theta_i + z\cos\theta_i)] && (2.92)\\
H_{y_1} &= \frac{E_0}{Z_2} T_\| \exp[-jk_2(x\sin\theta_t + z\cos\theta_t)] && (2.93)\\
H_{z_1} &= 0 && (2.94)\\
H_{z_2} &= 0 && (2.95)
\end{aligned}
$$

where

$$
\begin{aligned}
k_1 &= \omega\sqrt{\mu_1\varepsilon_1} && (2.96)\\
k_2 &= \omega\sqrt{\mu_2\varepsilon_2} && (2.97)\\
Z_1 &= \sqrt{\frac{\mu_1}{\varepsilon_1}} && (2.98)\\
Z_2 &= \sqrt{\frac{\mu_2}{\varepsilon_2}} && (2.99)
\end{aligned}
$$

and $T_\|$ and $\Gamma_\|$ are the paralell reflection and transmission coefficients, respectively.

For the boundary conditions, at $z = 0$ to be independent of x we must have

$$
\begin{aligned}
k_1 \sin\theta_i &= k_1 \sin\theta_r && (2.100)\\
k_1 \sin\theta_i &= k_2 \sin\theta_t && (2.101)
\end{aligned}
$$

From the above equation we get, if the media are lossless dielectrics

$$
\begin{aligned}
\theta_i &= \theta_r && (2.102)\\
\frac{\sin\theta_t}{\sin\theta_i} &= \frac{k_1}{k_2} && (2.103)\\
&= \sqrt{\frac{\mu_1\epsilon_1}{\mu_2\epsilon_2}} && (2.104)
\end{aligned}
$$

The reader will recognize in (2.102) and (2.103) Snell's laws of reflection and refraction.

If we assume further that $\mu_1 = \mu_2$, then (2.103) becomes simply

$$\frac{\sin\theta_t}{\sin\theta_i} = \sqrt{\frac{\epsilon_1}{\epsilon_2}} \tag{2.105}$$

The boundary conditions imply the continuity of the tangential components of the electric and the magnetic fields hence, for $z = 0$ we have, recalling (2.103),

$$(1 + \Gamma_{\|})\cos\theta_i = T_{\|}\cos\theta_t \tag{2.106}$$

$$(1 - \Gamma_{\|}) = T_{\|}\frac{Z_1}{Z_2} \tag{2.107}$$

from which we get the values of $\Gamma_{\|}$ and $T_{\|}$:

$$\Gamma_{\|} = \frac{Z_2\cos(\theta_t) - Z_1\cos(\theta_i)}{Z_2\cos(\theta_t) + Z_1\cos(\theta_i)} \tag{2.108}$$

$$T_{\|} = \frac{2Z_2\cos(\theta_t)}{Z_2\cos(\theta_t) + Z_1\cos(\theta_i)} \tag{2.109}$$

Similarly, the reflection and transmission coefficients for the electric field normal to the plane of incidence, that is, transverse electric, perpendicular, or horizontal polarization, $\Gamma_{\perp}$ and $T_{\perp}$, can be derived as

$$\Gamma_{\perp} = \frac{Z_2\sec(\theta_t) - Z_1\sec(\theta_i)}{Z_2\sec(\theta_t) + Z_1\sec(\theta_i)} \tag{2.110}$$

$$T_{\perp} = \frac{2Z_2\sec(\theta_t)}{Z_2\sec(\theta_t) + Z_1\sec(\theta_i)} \tag{2.111}$$

For an arbitrary polarization, the ratio of transmitted power density S_t to incident power density S_i is

$$\frac{S_t}{S_i} = \frac{|E_{\|_i}|^2 T_{\|}^2 + |E_{\perp_i}|^2 T_{\perp}^2}{|E_{\|_i}|^2 + |E_{\perp_i}|^2}\sqrt{\frac{\varepsilon_2}{\varepsilon_1}} \tag{2.112}$$

The power transmission coefficient, or transmissivity[4], the ratio of transmitted power density P_t across the boundary to the incident power density P_i is (see Figure 2.4)

$$\frac{P_t}{P_i} = \frac{S_t\,dA_t}{S_i\,dA_i} = 1 - |\Gamma|^2 \tag{2.113}$$

From Figure 2.4 we get

$$\frac{dA_t}{dA_i} = \frac{\cos\theta_t}{\cos\theta_i} \tag{2.114}$$

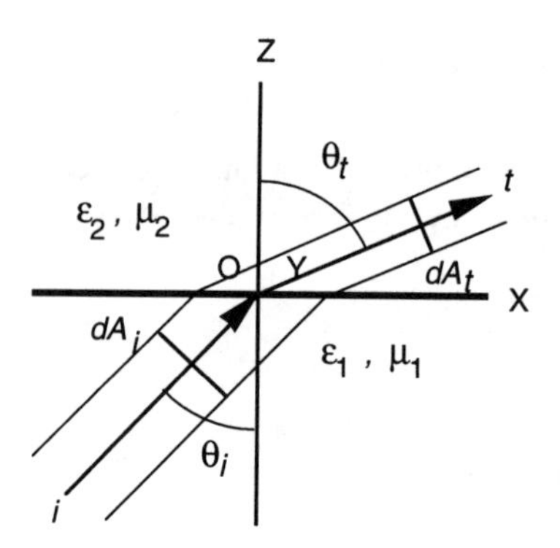

Figure 2.4: Relation between dA_i and dA_t for oblique incidence of a TEM wave on the plane boundary between two media.

Hence

$$\frac{P_t}{P_i} = \frac{S_t \cos\theta_t}{S_i \cos\theta_i} \tag{2.115}$$

2.9 EQUIVALENCE THEOREM

A given distribution of currents and charges originates a unique electromagnetic field. The inverse, however, is not true. Identical electromagnetic fields may be obtained by different source distributions. To solve some radiation problems it may be easier to substitute the original field sources with others that are equivalent, that is, that produce the same radiation fields in a given region of space.

Assume, as in Figure 2.5, that the field sources are contained in a limited volume V_1 enclosed by surface S. The volume outside S will be denoted by V_2. The field sources $(\boldsymbol{J}_e, \rho)$ originate a unique electromagnetic field $(\boldsymbol{E}, \boldsymbol{H})$ both inside and outside V_1. From the boundary conditions we know that the tangential components of $\boldsymbol{E}$ and $\boldsymbol{H}$ are continuous on S.

Consider now sources that give rise to fields $(\boldsymbol{E}_1, \boldsymbol{H}_1)$ inside V_1 and $(\boldsymbol{E}_2, \boldsymbol{H}_2)$ inside V_2. For these sources to be equivalent to the first ones, we must have everywhere in V_2

$$\boldsymbol{E}_2 = \boldsymbol{E} \tag{2.116}$$
$$\boldsymbol{H}_2 = \boldsymbol{H} \tag{2.117}$$

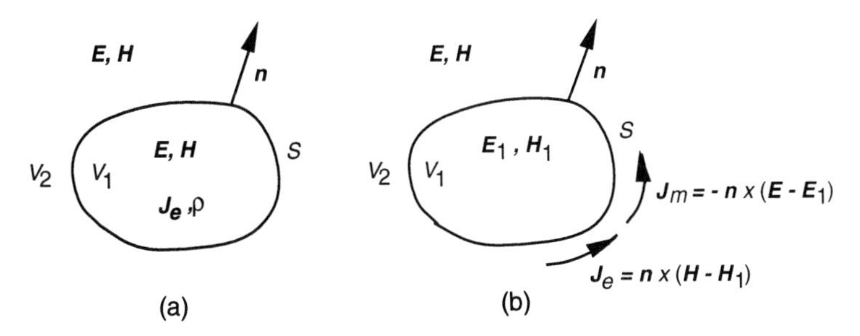

Figure 2.5: Equivalence theorem: (a) original sources and (b) equivalent sources.

The fields of the equivalent source must obey Maxwell's equations. Any discontinuity of the tangential components of the magnetic field on S must be compensated by an electric surface current density given by

$$\boldsymbol{J}_e = \boldsymbol{n} \times (\boldsymbol{H} - \boldsymbol{H}_1) \tag{2.118}$$

Similarly, any discontinuity of the tangential component of the electric field on S must give rise to a magnetic surface current density

$$\boldsymbol{J}_m = -\boldsymbol{n} \times (\boldsymbol{E} - \boldsymbol{E}_1) \tag{2.119}$$

The equivalence theorem states simply that the original current and charge distributions produce the same fields everywhere on V_2 as the surface current densities defined by (2.118) and (2.119). The equivalence theorem is also known as the Schelkunoff's equivalence principle and, in the special case where $\boldsymbol{E}_1 = 0$ and $\boldsymbol{H}_1 = 0$, it is known as Love's equivalence principle.

2.10 KIRCHHOFF-HUYGENS FORMULAS

In Section 2.7 we derived the wave equations in a source-free medium. Now we will look for solutions of Maxwell's equations in the presence of field sources (electric and magnetic charges and current densities).

Assuming a linear isotropic medium and a time dependence of the form $\exp(j\omega t)$, we show, in Appendix C, that the electromagnetic fields $\boldsymbol{E}_P$ and $\boldsymbol{H}_P$ at a point P, inside a finite volume V bounded by surface S, can be expressed in terms of the field sources existing inside volume V and the fields on S by

$$\boldsymbol{E}_P \;=\; -\frac{1}{4\pi}\int_V \left(j\omega\mu\psi\boldsymbol{J}_e + \boldsymbol{J}_m \times \boldsymbol{\nabla}\psi - \frac{\rho}{\varepsilon}\boldsymbol{\nabla}\psi\right)dV +$$

$$\frac{1}{4\pi}\int_S [-j\omega\mu\psi(\boldsymbol{n}\times\boldsymbol{H})+(\boldsymbol{n}\times\boldsymbol{E})\times\nabla\psi+(\boldsymbol{n}\cdot\boldsymbol{E})\nabla\psi]\,dS \tag{2.120}$$

$$\boldsymbol{H}_P = -\frac{1}{4\pi}\int_V\left(j\omega\varepsilon\psi\boldsymbol{J}_m-\boldsymbol{J}_e\times\nabla\psi-\frac{\rho_m}{\mu}\nabla\psi\right)dV + \frac{1}{4\pi}\int_S [j\omega\varepsilon\psi(\boldsymbol{n}\times\boldsymbol{E})+(\boldsymbol{n}\times\boldsymbol{H})\times\nabla\psi+(\boldsymbol{n}\cdot\boldsymbol{H})\nabla\psi]\,dS \tag{2.121}$$

where ψ is the Green function:

$$\psi = \frac{\exp(-jkr)}{r} \tag{2.122}$$

$\boldsymbol{n}$ is the unit vector normal at each point on the surface S, pointing toward V, r is the distance between point P and the general point Q inside V, and, as usual, $k^2=\omega^2\varepsilon\mu$.

Equations (2.120) and (2.121) are known as the Kirchhoff-Huygens integration formulas.

If volume V is unbounded, but the field sources are all contained in a finite volume, the integral over the outer surface boundary vanishes if the following conditions, also known as the radiation conditions, are met:

$$\lim_{R\to\infty} R\,|\boldsymbol{E}| \quad \text{is finite} \tag{2.123}$$

$$\lim_{R\to\infty} R\left|(\boldsymbol{r}\times\boldsymbol{H})+\sqrt{\frac{\varepsilon}{\mu}}\boldsymbol{E}\right| = 0 \tag{2.124}$$

$$\lim_{R\to\infty} R\,|\boldsymbol{H}| \quad \text{is finite} \tag{2.125}$$

$$\lim_{R\to\infty} R\left|\sqrt{\frac{\varepsilon}{\mu}}(\boldsymbol{r}\times\boldsymbol{E})-\boldsymbol{H}\right| = 0 \tag{2.126}$$

By an adequate choice of the boundary surfaces it is possible to exclude the field sources (charges and currents) from the volume V. If, in addition, this volume is outwardly limited by a sphere of infinite radius, and the fields obey the radiation conditions, we have

$$\boldsymbol{E}_P = \frac{1}{4\pi}\int_S [-j\omega\mu\psi(\boldsymbol{n}\times\boldsymbol{H})+(\boldsymbol{n}\times\boldsymbol{E})\times\nabla\psi+(\boldsymbol{n}\cdot\boldsymbol{E})\nabla\psi]\,dS \tag{2.127}$$

$$\boldsymbol{H}_P = \frac{1}{4\pi}\int_S [j\omega\varepsilon\psi(\boldsymbol{n}\times\boldsymbol{E})+(\boldsymbol{n}\times\boldsymbol{H})\times\nabla\psi+(\boldsymbol{n}\cdot\boldsymbol{H})\nabla\psi]\,dS \tag{2.128}$$

where S designates the closed surfaces that totally enclose the field sources.

In many cases of practical importance, however, it will be impossible (or extremely difficult) to specify the electric and magnetic field over a closed surface. In these cases it will be assumed that the contribution for integrals (2.127) and (2.128) will come mainly from a portion of surface S, which will be called the aperture A. This is in fact equivalent to assuming that everywhere over S, except on aperture A, the fields are zero or that $S - A$ is a perfectly conducting screen with neither currents nor charges.

In this case it is then necessary, in order to ensure the continuity of the electric and magnetic fields over A, to introduce distributions of electric and magnetic charges on the aperture line boundary (Γ_A). The effect of such a line of charges may be included by modifying the surface integrals as follows:

$$\boldsymbol{E}_P = \frac{1}{4\pi}\int_A \Bigg[-j\omega\mu\psi(\boldsymbol{n}\times\boldsymbol{H}) + (\boldsymbol{n}\times\boldsymbol{E})\times\nabla\psi + \frac{1}{j\omega\varepsilon}(\Sigma_j \boldsymbol{I}_j(\boldsymbol{n}\times\boldsymbol{H})\cdot\nabla\left(\frac{\partial\psi}{\partial x_j}\right)\Bigg] dS \tag{2.129}$$

$$\boldsymbol{H}_P = \frac{1}{4\pi}\int_A \Bigg[+j\omega\varepsilon\psi(\boldsymbol{n}\times\boldsymbol{E}) + (\boldsymbol{n}\times\boldsymbol{H})\times\nabla\psi - \frac{1}{j\omega\mu}(\Sigma_j \boldsymbol{I}_j(\boldsymbol{n}\times\boldsymbol{E})\cdot\nabla\left(\frac{\partial\psi}{\partial x_j}\right)\Bigg] dS \tag{2.130}$$

It should be noted, from (2.129) and (2.130), that only the tangential components of the electric and magnetic fields over the aperture are relevant to the radiation fields.

There is another formula, from Franz [5], to derive the radiated fields from the sources, which may be considered superior [6] to the one presented here in the near field. In the far field, however, both formulas are equivalent.

The Kirchhoff-Huygens formulas may be further simplified if we restrict the position of the observation point P to be far away from the aperture (far-field condition). Then we have

$$\boldsymbol{E}_P = \frac{-jk\exp(-jkR)}{4\pi R}\boldsymbol{R}_1 \times \int_A [\boldsymbol{n}\times\boldsymbol{E} - Z\boldsymbol{R}_1\times(\boldsymbol{n}\times\boldsymbol{H})]\exp(jk\boldsymbol{\rho}\cdot\boldsymbol{R}_1)\, dS \tag{2.131}$$

where $\boldsymbol{R}_1$ is the unit vector from the origin to the observation point, $\boldsymbol{\rho}$ is the vector from the origin to the integration point on the aperture, and R is the distance from the origin to the observation point.

In the far-field region it is not necessary to compute the magnetic field separately because it can easily be shown that

$$\lim_{kr\to\infty}(\boldsymbol{H}_P) = -\frac{\boldsymbol{E}_P}{Z}\times\boldsymbol{R}_1 \tag{2.132}$$

In vacuum $Z = Z_0 = \sqrt{\mu_0/\varepsilon_0}$.

2.11 RADIATION PATTERNS OF PLANE APERTURES

2.11.1 General Case

In the following we will be dealing exclusively with the far-field region. Thus radiation fields will be expressed in spherical coordinates, denoted by subscripts θ and ϕ, and the observation point P will be referenced by its spherical coordinates θ and ϕ.

$\boldsymbol{R}_1$ is given by

$$\boldsymbol{R}_1 = \sin(\theta)\cos(\phi)\boldsymbol{x} + \sin(\theta)\sin(\phi)\boldsymbol{y} + \cos(\theta)\boldsymbol{z} \tag{2.133}$$

If aperture A is restricted to be an equiphase surface, normal to the Z axis then (2.131) simplifies to

$$E_{P_\theta} = \frac{-jk\exp(-jkR)}{4\pi R}[\cos\phi(E_x + Z_0H_y\cos\theta) + \sin\phi(E_y - Z_0H_x\cos\theta)]$$
$$\exp[jk(x\sin\theta\cos\phi + y\sin\theta\sin\phi)]\,dS \tag{2.134}$$

$$E_{P_\phi} = \frac{-jk\exp(-jkR)}{4\pi R}[\cos\phi(-Z_0H_x + E_y\cos\theta) - \sin\phi(Z_0H_y + E_x\cos\theta)]$$
$$\exp[jk(x\sin\theta\cos\phi + y\sin\theta\sin\phi)]\,dS \tag{2.135}$$

If, in addition, on the aperture, $\boldsymbol{H}$ is related to $\boldsymbol{E}$ by

$$\boldsymbol{H} = \frac{\boldsymbol{z} \times \boldsymbol{E}}{Z_1} \tag{2.136}$$

with Z_1 being the aperture impedance, (2.134) and (2.135) further simplify to

$$E_{P_\theta} = \frac{jk\exp(-jkR)}{4\pi R}\left[1 + \cos(\theta)\frac{Z}{Z_1}\right][N_x\cos(\phi) + N_y\sin(\phi)] \tag{2.137}$$

$$E_{P_\phi} = \frac{-jk\exp(-jkR)}{4\pi R}\left[cos(\theta) + \frac{Z}{Z_1}\right][N_x\sin(\phi) - N_y\cos(\phi)] \tag{2.138}$$

where

$$N_x = \int_A E_x\exp[jk(x\sin\theta\cos\phi + y\sin\theta\sin\phi)]\,dS \tag{2.139}$$

$$N_y = \int_A E_y\exp[jk(x\sin\theta\cos\phi + y\sin\theta\sin\phi)]\,dS \tag{2.140}$$

The functions

$$E_\theta(\theta,\phi) = \frac{E_{P_\theta}(\theta,\phi)}{E_{P_\theta}(\theta_0,\phi_0)} \tag{2.141}$$

$$E_\phi(\theta,\phi) = \frac{E_{P_\phi}(\theta,\phi)}{E_{P_\phi}(\theta_0,\phi_0)} \tag{2.142}$$

where θ_0 and ϕ_0 are arbitrary values — usually boresight ($\theta_0 = 0$ and $\phi_0 = 0$) or the direction at which the radiated field is maximum — are known as the far-field radiation patterns.

For a rectangular aperture with dimensions a along the X axis and b along the Y axis, when the origin of the coordinate system is centered on the aperture, (2.139) and (2.140) become

$$N_x = \int_{-a/2}^{+a/2} dx \int_{-b/2}^{+b/2} E_x \exp[jk(x \sin\theta \cos\phi + y \sin\theta \sin\phi)]\, dy \quad (2.143)$$

$$N_y = \int_{-a/2}^{+a/2} dx \int_{-b/2}^{+b/2} E_y \exp[jk(x \sin\theta \cos\phi + y \sin\theta \sin\phi)]\, dy \quad (2.144)$$

For a plane circular aperture with radius ρ_1, centered on the origin of the coordinate system, (2.139) and (2.140) are modified as follows:

$$N_x = \int_0^{\rho_1} \rho\, d\rho \int_0^{2\pi} E_x \exp[jk\rho \sin\theta \cos(\phi - \varphi)]\, d\varphi \quad (2.145)$$

$$N_y = \int_0^{\rho_1} \rho\, d\rho \int_0^{2\pi} E_y \exp[jk\rho \sin\theta \cos(\phi - \varphi)]\, d\varphi \quad (2.146)$$

where ρ and φ are the current position variables on the aperture.

When the aperture is not strictly an equiphase surface but its deviations from it are small so that the Poynting vector over the aperture does not deviate much from the direction of the Z axis, we can still make use of (2.137) and (2.138), provided N_x and N_y are modified to include the phase distributions ϕ_x and ϕ_y associated with x and y components of the electric field as follows:

$$N_x = \int_A E_x \exp(j\phi_x) \exp[jk(x \sin\theta \cos\phi + y \sin\theta \sin\phi)]\, dS \quad (2.147)$$

$$N_y = \int_A E_y \exp(j\phi_y) \exp[jk(x \sin\theta \cos\phi + y \sin\theta \sin\phi)]\, dS \quad (2.148)$$

Modifying (2.143) and (2.144) or (2.145) and (2.146) to include a phase variation across the aperture is a straightforward operation.

The radiation patterns defined by (2.141) and (2.142) are three-dimensional (3D) surfaces, mostly with complex shapes. To simplify their graphic representation one often restricts each pattern to two lines corresponding to the intersection of the 3D surface with two orthogonal planes defined by $\phi = 0$ and $\phi = \pi/2$, aligned with the principal polarization directions of the aperture fields, also called the principal planes. When the aperture field is linearly polarized the intersection with the plane parallel to the E field is known as the E-plane pattern and the other intersection is known as the H-plane pattern.

Usually the E-and the H-plane patterns are normalized to some appropriate value,often the value on boresight or the maximum value.

$$E_\theta(\theta, \phi = 0) = \frac{1 + \cos(\theta)\frac{Z}{Z_1}}{1 + \cos(\theta_0)\frac{Z}{Z_1}} \frac{N_x(\theta, \phi = 0)}{N_x(\theta_0, \phi = 0)} \tag{2.149}$$

$$E_\theta(\theta, \phi = \pi/2) = \frac{1 + \cos(\theta)\frac{Z}{Z_1}}{1 + \cos(\theta_0)\frac{Z}{Z_1}} \frac{N_y(\theta, \phi = \pi/2)}{N_y(\theta_0, \phi = 0)} \tag{2.150}$$

$$E_\phi(\theta, \phi = 0) = \frac{cos(\theta) + \frac{Z}{Z_1}}{cos(\theta_0) + \frac{Z}{Z_1}} \frac{N_y(\theta, \phi = 0)}{N_y(\theta_0, \phi = 0)} \tag{2.151}$$

$$E_\phi(\theta, \phi = \pi/2) = \frac{cos(\theta) + \frac{Z}{Z_1}}{cos(\theta_0) + \frac{Z}{Z_1}} \frac{N_x(\theta, \phi = \pi/2)}{N_x(\theta_0, \phi = \pi/2)} \tag{2.152}$$

Numeric evaluation of surface integrals in the expressions that define N_x and N_y, due to the oscillatory nature of the exponential integrand, presents considerable computational difficulties when the aperture dimensions are large compared to the wavelength. Salema [7] put forward a numerically efficient algorithm for such cases, applicable to both rectangular and circular large (and very large) apertures.

In some cases of practical importance, it is possible to make a few assumptions that simplify calculations, transforming the double integrals into single integrals.

To begin with,consider a rectangular aperture with dimensions a and b along the X axis and the Y axis, respectively, on which the aperture field components can be written as

$$E_x = f_x(x)\, f_y(y) \tag{2.153}$$

$$E_y = g_x(x)\, g_y(y) \tag{2.154}$$

Then $N_x(\theta, \phi)$ and $N_y(\theta, \phi)$ turn out as

$$N_x(\theta, \phi) = \int_{-a/2}^{+a/2} f_x(x) \exp[jkx \sin(\theta) \cos(\phi)]\, dx \int_{-b/2}^{+b/2} f_y(y) \exp[jky \sin(\theta) \sin(\phi)]\, dy \tag{2.155}$$

$$N_y(\theta, \phi) = \int_{-a/2}^{+a/2} g_x(x) \exp[jkx \sin(\theta) \cos(\phi)]\, dx \int_{-b/2}^{+b/2} g_y(y) \exp[jky \sin(\theta) \sin(\phi)]\, dy \tag{2.156}$$

From the above we notice that $N_x(\theta, \phi = 0)$ and $N_y(\theta, \phi = 0)$ depend on θ only through the integral on x. Similarly $N_x(\theta, \phi = \pi/2)$ and $N_y(\theta, \phi = \pi/2)$ depend on

θ only through the integral on y. If we now recall (2.149) to (2.152), we arrive at the following very important consequence: the radiation pattern in any of the principal planes is simply a function of the the aperture field along that plane. The variation of the aperture field along the other plane affects the radiated field only as a constant (independent of θ) multiplication factor and does not interfere with the radiation pattern.

In the following subsections we will apply the previously derived equations to a few simple cases of practical importance.

2.11.2 Rectangular Apertures

Take a plane rectangular aperture with a uniform illumination, linearly polarized along the Y axis, that is, $g_x(x) = 0$ and $g_y(y) = 1$. The aperture impedance will be assumed to be equal to the free space medium where the aperture is located ($Z_1 = Z_0$). Substituting in (2.156) we get

$$\begin{aligned} N_x(\theta, \phi) &= 0 \\ N_y(\theta, \phi) &= ab \frac{\sin\left[k\frac{a}{2}\sin(\theta)\cos(\phi)\right]}{k\frac{a}{2}\sin(\theta)\cos(\phi)} \frac{\sin\left[k\frac{b}{2}\sin(\theta)\sin(\phi)\right]}{k\frac{b}{2}\sin(\theta)\sin(\phi)} \end{aligned} \tag{2.157}$$

which yield the following radiation patterns in the principal planes, normalized to boresight ($\theta_0 = 0$):

$$\begin{aligned} E_\theta(\theta, \phi = 0) &= 0 \\ E_\phi(\theta, \phi = 0) &= \frac{1 + \cos\theta}{2} \frac{\sin\left(k\frac{a}{2}\sin\theta\right)}{k\frac{a}{2}\sin\theta} \end{aligned} \tag{2.158}$$

$$\begin{aligned} E_\theta(\theta, \phi = \frac{\pi}{2}) &= \frac{1 + \cos\theta}{2} \frac{\sin\left(k\frac{b}{2}\sin\theta\right)}{k\frac{b}{2}\sin\theta} \\ E_\phi(\theta, \phi = \frac{\pi}{2}) &= 0 \end{aligned} \tag{2.159}$$

The above equations indicate that the radiated field is linearly polarized and that the radiation pattern in each principal plane is only dependent on the aperture dimension along that plane. In Figure 2.6 we represent $20\log_{10}\left(\frac{\sin u}{u}\right)$ as a function of u. As shown, this function has a well-defined main lobe and a set of sidelobes with decreasing amplitudes. For large apertures ($ka >> 1$) the first sidelobe located at $u = 4.493$ is –13.3 dB below boresight level. For smaller apertures, due to the effect of the $1 + \cos\theta$ factor, the first sidelobe level may be considerably reduced.

Using (2.137) or (2.27) for boresight ($\theta = 0$) we obtain the boresight directivity (i.e., the maximum directivity) di_{unif} of the uniformly illuminated rectangular aperture, compared with the isotropic source:

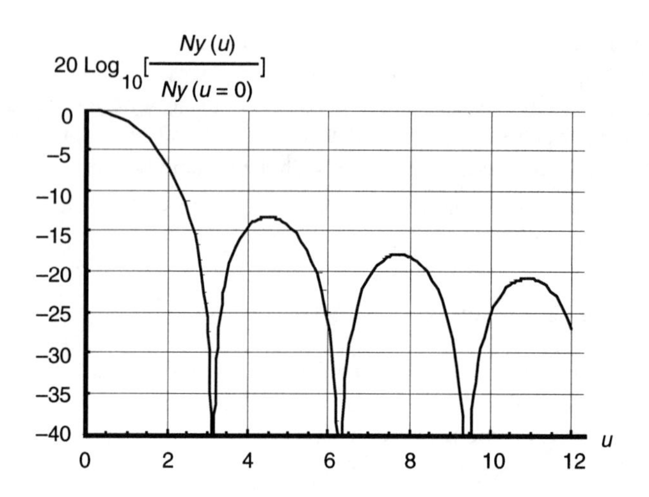

Figure 2.6: Far-field radiation pattern of a rectangular aperture with uniform illumination, omitting the factor $1 + \cos\theta$, as a function of $u = \frac{ka}{2}\sin\theta$, in the plane where a is the aperture size.

$$
\begin{aligned}
di_{\text{unif}} &= \frac{E^2_{P_\theta}(\theta = 0, \phi = 0)}{\frac{ab}{4\pi R^2}} \\
&= \frac{E^2_{P_\phi}(\theta = 0, \phi = \pi/2)}{\frac{ab}{4\pi R^2}} \\
&= \frac{4\pi}{\lambda^2} ab \qquad (2.160)
\end{aligned}
$$

Another case — which is approximately applicable to solid dielectric antennas with rectangular cross section — is the linearly polarized cosine-cosine illumination where

$$
\begin{aligned}
f_x(x) &= 0 \\
f_y(y) &= 0 \\
g_x(x) &= \cos(k_x x) \\
g_y(y) &= \cos(k_y y)
\end{aligned}
$$

k_x and k_y being arbitrary real constants. After considerable manipulation we get

$$
\begin{aligned}
N_x &= 0 \\
N_y &= 4\frac{k_x \cos(\frac{\chi a}{2})\sin(\frac{k_x a}{2}) - \chi\cos(\frac{k_x a}{2})\sin(\frac{\chi a}{2})}{k_x^2 - \chi^2} \times
\end{aligned}
$$

$$\frac{k_y \cos(\frac{\xi b}{2}) \sin(\frac{k_y b}{2}) - \xi \cos(\frac{k_y b}{2}) \sin(\frac{\xi b}{2})}{k_y^2 - \xi^2} \tag{2.161}$$

where

$$\chi = k \sin(\theta) \cos(\phi) \tag{2.162}$$
$$\xi = k \cos(\theta) \sin(\phi) \tag{2.163}$$

The unnormalized radiation pattern in the principal planes becomes

$$E_\theta(\theta, \phi = 0) = 0 \tag{2.164}$$

$$E_\phi(\theta, \phi = 0) = 4(1 + \cos\theta) \frac{\sin\left(\frac{k_y b}{2}\right)}{k_y} \times \frac{k_x \cos(u) \sin\left(\frac{k_x a}{2}\right) - k \sin\theta \cos\left(\frac{a k_x}{2}\right) \sin(u)}{k_x^2 - k^2 \sin^2\theta} \tag{2.165}$$

$$E_\theta(\theta, \phi = \frac{\pi}{2}) = 4(1 + \cos\theta) \frac{\sin\left(\frac{k_x a}{2}\right)}{k_x} \times \frac{k_y \cos(v) \sin\left(\frac{k_y b}{2}\right)) - k \sin\theta \cos\left(\frac{b k_y}{2}\right) \sin(v)}{k_y^2 - k^2 \sin^2\theta} \tag{2.166}$$

$$E_\phi(\theta, \phi = \frac{\pi}{2}) = 0 \tag{2.167}$$

where

$$u = \frac{ka}{2} \sin\theta \tag{2.168}$$
$$v = \frac{kb}{2} \sin\theta \tag{2.169}$$

Here again we find that the H-plane radiation pattern, except for a constant multiplying factor, does not vary with k_y and that the E-plane radiation pattern is independent of k_x.

Proceeding as above we get the maximum directivity (boresight directivity) of the rectangular aperture with cosine illumination $di_{\cos}$:

$$di_{\cos} = \frac{4\pi}{\lambda^2} ab \left[\frac{2 \sin(\frac{k_x a}{2}) \sin(\frac{k_y b}{2})}{(\frac{k_x a}{2})(\frac{k_y b}{2})} \right]^2 \tag{2.170}$$

A particularly interesting case for dielectric antennas with rectangular cross section occurs when

$$\frac{k_x a}{2} = \frac{\pi}{2}$$
$$\frac{k_y b}{2} = \frac{\pi}{2}$$

for which the radiation pattern in the principal planes simplifies to

$$E_\theta(\theta, \phi = 0) = 0$$
$$E_\phi(\theta, \phi = 0) = (1 + \cos\theta)\frac{4b}{\pi}\frac{\frac{\pi}{a}\cos\left(\frac{ka\sin\theta}{2}\right)}{\frac{\pi^2}{a^2} - k^2\sin^2\theta} \quad (2.171)$$

$$E_\theta(\theta, \phi = \frac{\pi}{2}) = (1 + \cos\theta)\frac{4a}{\pi}\frac{\frac{\pi}{b}\cos\left(\frac{kb\sin\theta}{2}\right)}{\frac{\pi^2}{b^2} - k^2\sin^2\theta} \quad (2.172)$$
$$E_\phi(\theta, \phi = \frac{\pi}{2}) = 0$$

and the directivity on boresight becomes

$$di_{\cos} = \frac{4\pi}{\lambda^2}\, ab\, \frac{64}{\pi^4} \quad (2.173)$$

representing an aperture illumination efficiency $\eta = \frac{64}{\pi^4}$ or $\eta \approx 0.657$.

In Figure 2.7 we plot $20\log_{10}\left(\frac{\cos u}{1-\frac{4u^2}{\pi^2}}\right)$ as a function of u, which is directly comparable to the function represented in Figure 2.6. As above the radiation pattern has a well-defined main lobe and a set of sidelobes with decreasing amplitudes. Compare this pattern with that for uniform illumination and you will immediately recognize the pattern widening (consequence of the lower aperture efficiency) as well as the lower sidelobe levels. For large apertures ($ka >> 1$) the first sidelobe now appears at $u = 5.936$ and its value is -23 dB, almost 10 dB less than for uniform illumination.

Taking these first two examples, of uniform and cosine-cosine illumination, let us evaluate the effect of the aperture departing from an equiphase surface under the conditions applicable to (2.147) and (2.148).

If ϕ_x and ϕ_y are linear functions of aperture coordinates x and y then it is simple to show that the radiation pattern is unaffected except for a shift in the direction of the maximum radiated field that ceases to be boresight ($\theta = 0, \phi = 0$).

If ϕ_x and ϕ_y are quadratic functions of aperture coordinates x and y, the radiation patterns change in three ways:

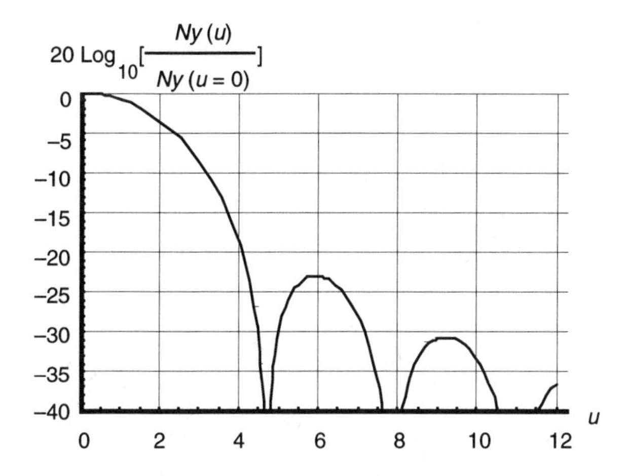

Figure 2.7: Far-field radiation pattern of a rectangular aperture with cosine illumination, omitting the factor $1 + \cos\theta$, as a function of $u = \frac{ka}{2}\sin\theta$, in the plane where a is the aperture size.

- Maximum directivity (and aperture efficiency) decreases;
- Pattern nulls are filled in;
- Sidelobe levels increase.

The magnitude of these effects depends on the type of aperture illumination and in general decreases as the aperture fields taper down toward the aperture edges.

Let us take a uniform amplitude linearly polarized along the Y axis aperture field and a quadratic phase error along X axis as follows:

$$\phi_y = \frac{4\Phi}{a^2}x^2 \tag{2.174}$$

where Φ is the maximum phase error, in radians, at the aperture edge ($x = \pm a/2$).

Substituting (2.174) in (2.148) and integrating we get

$$\begin{aligned} N_y &= \frac{ab\sqrt{\pi}}{4\sqrt{\Phi}}(-1)^{3/4}\exp(\frac{ju^2}{4\Phi}) \\ &\left\{\text{erf}\left[\frac{(-1)^{1/4}(\text{u} - 2\Phi)}{2\sqrt{\Phi}}\right] - \text{erf}\left[\frac{(-1)^{1/4}(\text{u} + 2\Phi)}{2\sqrt{\Phi}}\right]\right\} \end{aligned} \tag{2.175}$$

where, as above, $u = \frac{ka}{2}\sin\theta$.

In Figure 2.8 we plot $20\log_{10}\left(\frac{N_y}{ab}\right)$ as a function of u for $\Phi = \pi/4$, $\Phi = \pi/2$ and $\Phi = \pi$. To make comparison with Figure 2.6 easier we also represent in Figure 2.8 the curve for $\Phi = 0$.

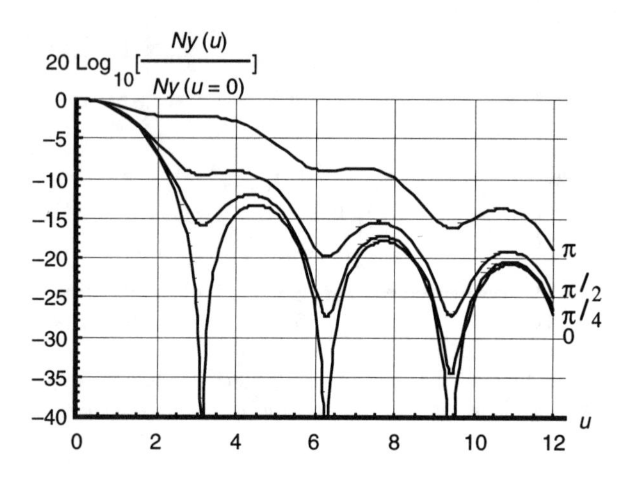

Figure 2.8: Far-field radiation pattern of a rectangular aperture with uniform illumination and quadratic phase error — omitting the factor $1+\cos\theta$ — as a function of $u = \frac{ka}{2}\sin\theta$, in the plane where a is the aperture size, for maximum phase error at the edge of 0, $\pi/4$, $\pi/2$, and π.

By making use of Figure 2.8 we may easily quantify the effects of a quadratic phase error. For small values of the maximum phase error the main effect is null filling, particularly in the first null. For increasing maximum phase errors the other effects — namely, increasing sidelobe levels and decreasing boresight directivity — become more apparent.

Figure 2.9 represents boresight directivity loss Δ_Φ versus phase error Φ at the aperture edge and may be used, in conjunction with (2.160), to calculate the directivity of a rectangular aperture with uniform illumination and quadratic phase error along one of the aperture axis. If the quadratic phase error occurs along both aperture axes then the directivity loss given in Figure 2.9 simply has to be applied twice, once for the phase error at each aperture edge. For a rectangular aperture of sizes a and b, with phase errors Φ_x at $y = 0$ and $x = \pm a/2$ and Φ_y at $x = 0$ and $y = \pm b/2$ boresight directivity in dBi is

$$10\log_{10}\left(\frac{4\pi}{\lambda^2}\,ab\right) - \Delta_{\Phi_x} - \Delta_{\Phi_y} \tag{2.176}$$

For cosine illumination with a quadratic phase error the integral expressions in N_y are somewhat more involved but may still be performed in close form:

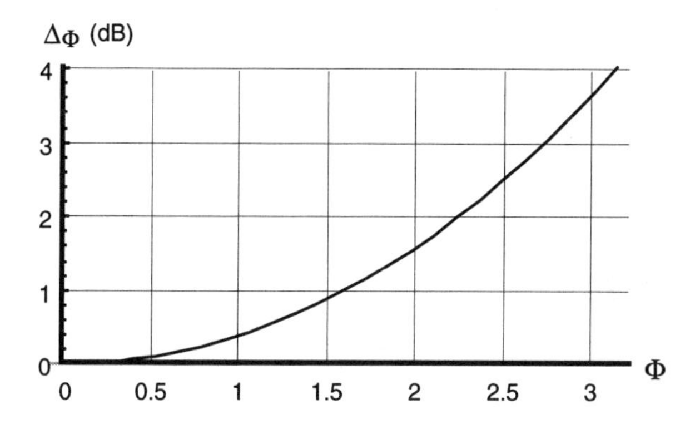

Figure 2.9: Boresight directivity loss Δ_Φ for a rectangular aperture with uniform illumination with quadratic phase error along one dimension versus maximum phase error Φ at the edge.

$$
\begin{aligned}
N_y &= -\frac{ab}{4\sqrt{\pi}\sqrt{\Phi}}(-1)^{3/4}\exp(\frac{j(\pi-2u)^2}{16\Phi}) \\
&\left\{\mathrm{erf}\left[\frac{(-1)^{1/4}(\frac{\pi}{2}-u+2\Phi)}{2\sqrt{\Phi}}\right]+\mathrm{erf}\left[\frac{(-1)^{1/4}(-\frac{\pi}{2}+u+2\Phi)}{2\sqrt{\Phi}}\right]-\exp(\frac{j\pi u}{2\Phi})\times\right. \\
&\left.\left\{\mathrm{erf}\left[\frac{(-1)^{1/4}(\frac{\pi}{2}+u-2\Phi)}{2\sqrt{\Phi}}\right]-\mathrm{erf}\left[\frac{(-1)^{1/4}(\frac{\pi}{2}+u+2\Phi)}{2\sqrt{\Phi}}\right]\right\}\right\} \qquad (2.177)
\end{aligned}
$$

where, as before, $u = \frac{ka}{2}\sin\theta$.

Figure 2.10 represents $20\log_{10}\left(\frac{N_y}{ab}\right)$ as a function of u for $\Phi = \pi/4$, $\Phi = \pi/2$, and $\Phi = \pi$. For easy comparison with the uniform phase case, we also represent in Figure 2.10 the curve for $\Phi = 0$.

The same remarks made about the uniform illumination with a quadratic phase error apply here. Comparing Figures 2.8 and 2.10 we note that, for the same maximum phase error, boresight directivity loss is somewhat smaller for the cosine illumination, whereas the null filling and the sidelobe level increase are slightly larger. In fact, the combination of the last two effects could make the radiation pattern appear to be sidelobe-free.

As for the uniform illumination case we represent in Figure 2.11 boresight directivity loss Δ_Φ versus phase error at the aperture edge Φ for cosine illumination. Figure 2.11 may be used in conjunction with (2.173) to calculate the directivity of a rectangular aperture with cosine illumination and quadratic phase error along one aperture axis. For quadratic phase errors along both aperture axes the previous remarks also apply here.

If we recall that the transversal electric field of the $TE_{1,0}$ mode in a metallic rectangular waveguide is uniform along the E-plane and follows a cosine law along the H-plane,

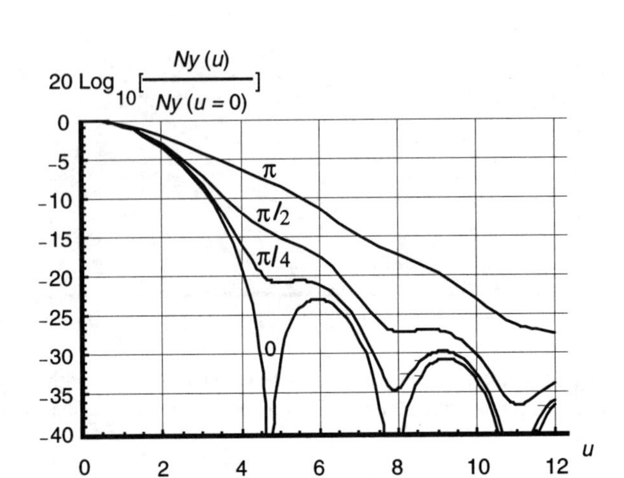

Figure 2.10: Far-field radiation pattern of a rectangular aperture with cosine illumination and quadratic phase error — omitting the factor $1 + \cos\theta$ — as a function of $u = \frac{ka}{2}\sin\theta$, in the plane where a is the aperture size, for maximum phase error at the edge of 0, $\pi/4$, $\pi/2$, and π.

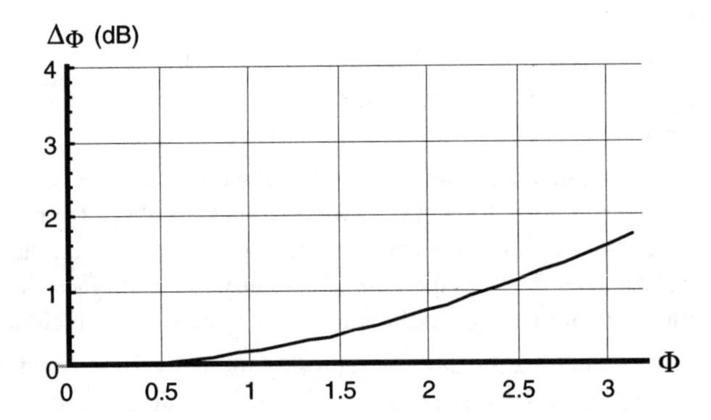

Figure 2.11: Boresight directivity loss Δ_Φ for a rectangular aperture with cosine illumination with quadratic phase error along one dimension versus maximum phase error Φ at the edge.

boresight directivity and radiation patterns along the principal planes of metallic rectangular waveguides and pyramidal horns excited in this mode may be easily estimated using Figures 2.6 through 2.11 with maximum phase errors at the aperture edges:

$$\Phi_E = \frac{\pi}{\lambda}\frac{b^2}{4r_E} \tag{2.178}$$

$$\Phi_H = \frac{\pi}{\lambda}\frac{a^2}{4r_H} \tag{2.179}$$

where r_E and r_H are the radius of curvature of the horn aperture, respectively, along the E- and the H-plane.

As mentioned above, when using Figures 2.8 or 2.10 to obtain the radiation pattern of a given aperture, one should not forget to multiply the values represented by the factor $1 + \cos\theta$.

2.11.3 Circular Apertures

Assume now a plane circular aperture, with radius ρ_1 centered on the origin of the coordinate system and aperture fields E_x and E_y of the form

$$E_x = E_x(\rho)\sin(n\varphi) \tag{2.180}$$

$$E_y = E_y(\rho)\cos(n\varphi) \tag{2.181}$$

where ρ and φ are the general polar coordinates on the aperture and n is an integer. The aperture impedance will again be taken equal to the free space medium where the aperture is located ($Z_1 = Z_0$)

In this case after considerable manipulation we arrive at

$$N_x = 2\pi j^n \sin(n\phi)\int_0^{\rho_1} E_x(\rho)J_n(k\rho\sin\theta)\rho\, d\rho \tag{2.182}$$

$$N_y = 2\pi j^n \cos(n\phi)\int_0^{\rho_1} E_y(\rho)J_n(k\rho\sin\theta)\rho\, d\rho \tag{2.183}$$

where $J_n(x)$ is the Bessel function of the first kind, order n and argument x, and ρ_1 is the aperture radius.

We will now apply (2.182) to a few cases of practical interest. Take, to start, a linearly polarized (along the Y axis), uniform illumination where

$$E_y(\rho) = 1 \tag{2.184}$$

$$n = 0 \tag{2.185}$$

Substituting in (2.183) and manipulating we get

$$\begin{aligned} N_y &= 2\pi\rho_1^2 \int_0^1 J_0(ux)x dx \\ &= 2\pi\rho_1^2 \frac{J_1(ux)}{u} \end{aligned} \tag{2.186}$$

where

$$u = k\rho_1 \sin\theta \tag{2.187}$$

From the above we get the following radiation patterns in the principal planes, normalized to boresight ($\theta_0 = 0$):

$$\begin{aligned} E_\theta(\theta, \phi = 0) &= 0 \\ E_\phi(\theta, \phi = 0) &= (1 + \cos\theta)\frac{J_1(k\rho_1 \sin\theta)}{k\rho_1 \sin\theta} \end{aligned} \tag{2.188}$$

$$\begin{aligned} E_\theta(\theta, \phi = \frac{\pi}{2}) &= (1 + \cos\theta)\frac{J_1(k\rho_1 \sin\theta)}{k\rho_1 \sin\theta} \\ E_\phi(\theta, \phi = \frac{\pi}{2}) &= 0 \end{aligned} \tag{2.189}$$

The radiated field is linearly polarized and the radiation pattern is circular symmetric, that is, identical in both principal planes. As shown in Figure 2.12, which represents $20\log_{10}\left(\frac{J_1(u)}{u}\right)$ as a function of u, the radiation pattern is very similar to the one for the rectangular aperture, with a well-defined main lobe and a set of sidelobes with decreasing amplitudes. For large apertures ($k\rho_1 >> 1$) the first sidelobe located at $u = 5.1356$ is -17.6 dB below boresight level, that is, 4.3 dB less than for the rectangular aperture. As we noted above, for smaller apertures the first sidelobe level may be considerably reduced, due to the effect of the $1 + \cos\theta$ factor.

Using (2.137) or (2.138) for boresight ($\theta = 0$) we obtain boresight directivity (i.e., the maximum directivity) di_{unif} compared with an isotropic source:

$$\begin{aligned} di_{\text{unif}} &= \frac{E_{P_\theta}^2(\theta = 0, \phi = 0)}{\frac{\pi\rho_1^2}{4\pi R^2}} \\ &= \frac{E_{P_\phi}^2(\theta = 0, \phi = \pi/2)}{\frac{\pi\rho_1^2}{4\pi R^2}} \\ &= \frac{4\pi}{\lambda^2}\pi\rho_1^2 \\ &= (k\rho_1)^2 \end{aligned} \tag{2.190}$$

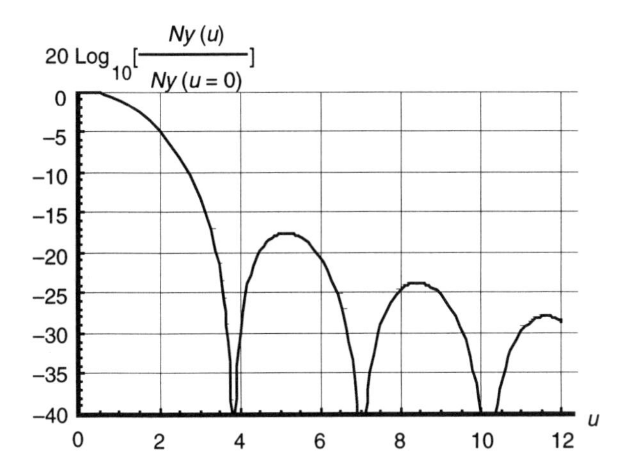

Figure 2.12: Far-field radiation pattern of a circular aperture with uniform illumination, omitting the factor $1 + \cos\theta$, as a function of $u = k\rho_1 \sin\theta$.

Another case, which is approximately applicable to solid dielectric antennas with circular cross section, is the linearly polarized cosine illumination where

$$
\begin{aligned}
E_y(\rho) &= \cos\left(\frac{\pi}{2}\frac{\rho}{\rho_1}\right) \\
n &= 0
\end{aligned}
$$

Substituting in (2.182) and manipulating we get

$$N_y = 2\pi\rho_1^2 \int_0^1 x \cos(\frac{\pi}{2}x) J_0(ux) dx \tag{2.191}$$

where, as before, $u = k\rho_1 \sin\theta$.

The unnormalized radiation pattern in the principal planes becomes

$$
\begin{aligned}
E_\theta(\theta, \phi = 0) &= 0 \\
E_\phi(\theta, \phi = 0) &= (1 + \cos\theta) 2\pi\rho_1^2 \int_0^1 x \cos(\frac{\pi}{2}x) J_0(ux) dx && (2.192) \\
E_\theta(\theta, \phi = \frac{\pi}{2}) &= (1 + \cos\theta) 2\pi\rho_1^2 \int_0^1 x \cos(\frac{\pi}{2}x) J_0(ux) dx && (2.193) \\
E_\phi(\theta, \phi = \frac{\pi}{2}) &= 0
\end{aligned}
$$

Noting that the power radiated by the circular aperture is

$$
\begin{aligned}
p_{\cos} &= \frac{1}{Z_0}\int_0^{2\pi} d\varphi \int_0^{\rho_1} \cos^2\left(\frac{\pi}{2}\frac{\rho}{\rho_1}\right) d\rho \\
&= \frac{\pi\rho_1^2}{Z_0} 2\left(\frac{1}{4} - \frac{1}{\pi^2}\right)
\end{aligned}
\tag{2.194}
$$

we get the maximum directivity (boresight directivity) of the circular aperture with cosine illumination $di_{\cos}$ compared to the isotropic source radiating the same power:

$$
\begin{aligned}
di_{\cos} &= \frac{4\pi}{\lambda^2}\pi\rho_1^2 \frac{2\left(\frac{2}{\pi} - \frac{4}{\pi^2}\right)^2}{\frac{1}{4} - \frac{1}{\pi^2}} \\
&\approx 0.72(k\rho_1)^2
\end{aligned}
\tag{2.195}
$$

which leads to an aperture illumination efficiency $eta \approx 0.72$.

Figure 2.13 represents $20\log_{10}\left(\frac{N_x(u)}{N_x(0)}\right)$ as a function of u. Comparing it with Figure 2.12 we notice at once for the cosine illumination that there is widening of the main lobe, a consequence of lower aperture efficiency, and reduced sidelobe levels . For large apertures ($k\rho_1 >> 1$) the first sidelobe now appears at $u = 6.536$ and its value at -26.1 dB is almost 9 dB less than for the case of uniform illumination.

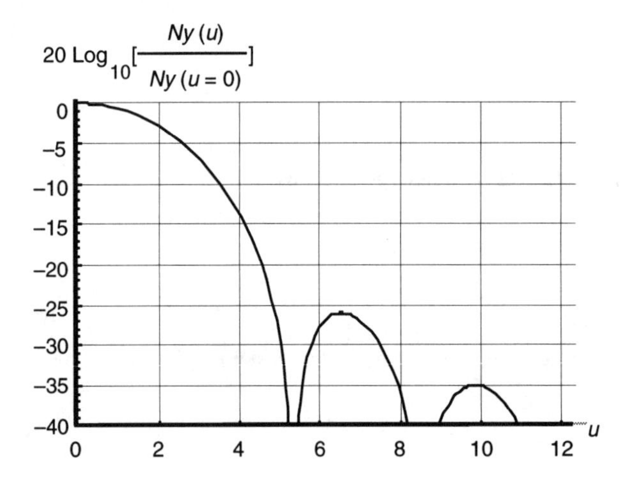

Figure 2.13: Far-field radiation pattern of a circular aperture with cosine illumination, omitting the factor $1 + \cos\theta$, as a function of $u = k\rho_1 \sin\theta$.

For later reference it is important to look into the effects of phase errors. In the following we will assume the conditions applicable to (2.147) and (2.148).

Linear phase errors simply result in a shift in the direction of the maximum radiated field which ceases to be boresight ($\theta = 0$, $\phi = 0$). If the phase error ϕ_ρ is a quadratic function of the aperture coordinate ρ the radiation patterns change in three ways:

- Maximum directivity (and aperture efficiency) decreases;
- Pattern nulls are filled in;
- Sidelobe levels increase.

The magnitude of these effects depends on the type of aperture illumination and in general decreases as the aperture fields taper down towards the aperture edges. In order to quantify these effects we will take in turn the uniform and the cosine illumination.

Let us take a radial quadratic phase error on the aperture

$$\phi_\rho = -\Phi \left(\frac{\rho}{\rho_1} \right)^2 \tag{2.196}$$

where Φ is the maximum phase error, in radians, at the aperture edge ($\rho = \rho_1$).

Introducing the quadratic phase error, the field expressions for the uniform and the cosine illumination become, respectively,

$$E_x(\rho) = \exp\left[-j \left(\frac{\rho}{\rho_1} \right)^2 \right] \tag{2.197}$$

$$E_x(\rho) = \cos\left(\frac{\pi}{2} \frac{\rho}{\rho_1} \right) \exp\left[-j \left(\frac{\rho}{\rho_1} \right)^2 \right] \tag{2.198}$$

which lead to, for the uniform illumination

$$N_y = 2\pi\rho_1^2 \int_0^1 x J_0(ux) \exp(-j\Phi x^2) dx \tag{2.199}$$

and for the cosine illumination

$$N_y = 2\pi\rho_1^2 \int_0^1 x J_0(ux) \cos\left(\frac{\pi}{2} x \right) \exp(-j\Phi x^2) dx \tag{2.200}$$

where, as before, $u = k\rho_1 \sin\theta$.

Figure 2.14 represents $20\log_{10}\left(\frac{N_y}{N_y(u=0)} \right)$ as a function of u for $\Phi = \pi/4, \Phi = \pi/2$, and $\Phi = \pi$. To easy comparison with Figure 2.12 we also represent in 2.14 the curve for $\Phi = 0$.

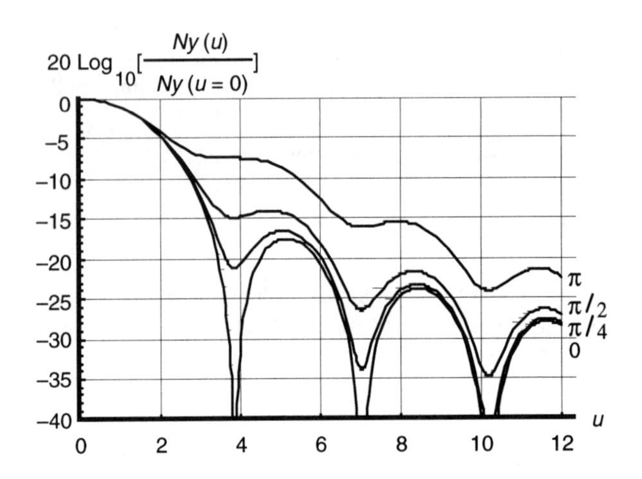

Figure 2.14: Far-field radiation pattern of a plane circular aperture with uniform illumination and quadratic phase error, omitting the factor $1 + \cos\theta$, as a function of $u = k\rho \sin\theta$, for maximum phase error at the edge of 0, $\pi/4$, $\pi/2$, and π.

Looking at Figure 2.14 we identify the same phenomena which were already referred to for the rectangular aperture. For small values of the phase error at the edge the main effect is null filling, particularly noticeable in the first null. For increasing values of the phase the other effects, namely, increasing sidelobe levels and decreasing boresight directivity, become more apparent.

Figure 2.15 represents boresight directivity loss Δ_Φ versus phase error at the aperture edge Φ and may be used, in conjunction with (2.190) to calculate the directivity of a plane circular aperture with uniform illumination and quadratic phase error along the radius, as follows:

$$di_{\text{dB}} = 10\log_{10}\left(k\rho_1^2\right) - \Delta_\Phi \tag{2.201}$$

Function $20\log_{10}\left(\frac{N_y}{N_y(u=0)}\right)$ as a function of u for $\Phi = 0$, $\Phi = \pi/4$, $\Phi = \pi/2$, and $\Phi = \pi$ is represented in Figure 2.16.

We represent in Figure 2.17 boresight directivity loss Δ_Φ versus phase error at the aperture edge Φ for the cosine illumination.

Comparing Figures 2.9 and 2.15 and Figures 2.14 and 2.16 we note that, for the same maximum phase error, boresight directivity loss is somewhat smaller for cosine illumination whereas the increase in the null filling and sidelobe level is slightly larger. The combination of the last two effects may make the radiation pattern appear to be sidelobe-free.

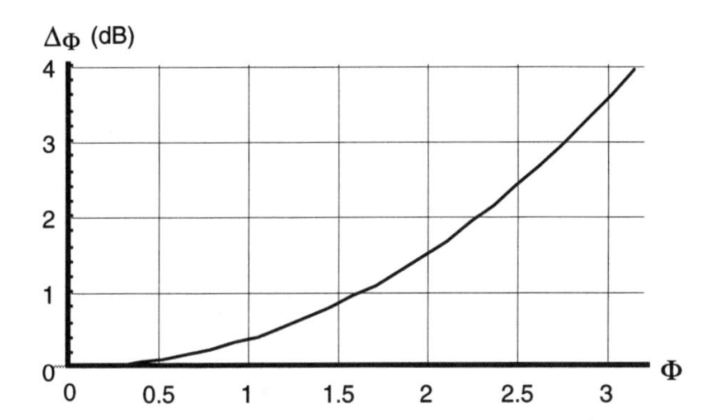

Figure 2.15: Boresight directivity loss Δ_Φ for a circular aperture with uniform illumination and quadratic phase error along the radius versus phase error Φ at the edge.

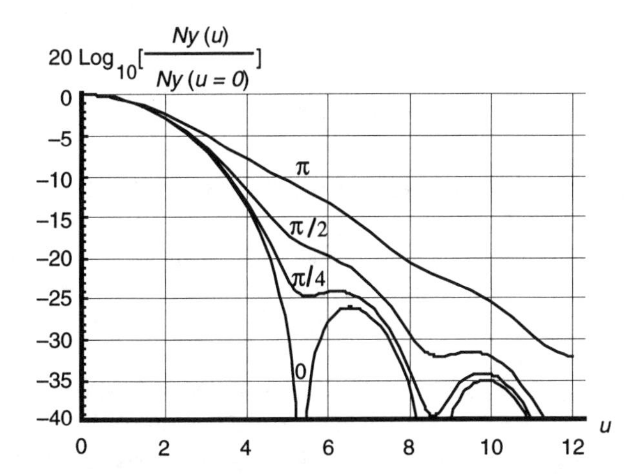

Figure 2.16: Far-field radiation pattern of a circular aperture with cosine illumination and quadratic phase error, omitting the factor $1+\cos\theta$, as a function of $u = k\rho\sin\theta$, for phase error at the edge of 0, $\pi/4$, $\pi/2$, and π.

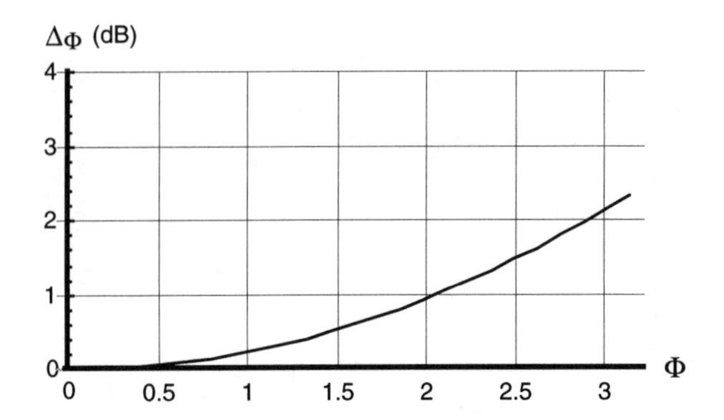

Figure 2.17: Boresight directivity loss Δ_Φ for a circular aperture with cosine illumination with quadratic phase error versus phase error Φ at the edge.

The transversal electric field of the $TE_{1,1}$ spherical mode of a conical metallic horn may be roughly taken as constant along the E-plane and as following a cosine law along the H-plane. Thus boresight directivity and radiation patterns along the principal planes of such antennas may be estimated using Figures 2.12 to 2.17 where the phase error at the edge is given by

$$\Phi = \frac{\pi \rho_1^2}{\lambda r} \tag{2.202}$$

and r is the radius of curvature of the horn aperture.

2.12 RADIATION PATTERNS OF SPHERICAL CAP APERTURES

Even if, for simplicity, a plane aperture is used whenever possible, there are cases where aperture phase errors are such that the Poynting vector can no longer be assumed to be coincident with the Z axis and the general form of the Kirchhoff-Huygens formulas must be used.

Sometimes, for example when dealing with conical horns, we may specify a spherical cap as the equiphase aperture, for which a particular form of the Kirchhoff-Huygens formulas may be derived. Consider a cap obtained from a sphere centered in the origin of the coordinate system, with radius R_1 and delimited by an angle θ_1, measured from the center. In this case, if the aperture fields expressed in spherical coordinates have unity azimuthal dependence, that is, if they are of the form

$$E_\vartheta = E_\vartheta(\vartheta)\sin(\varphi) \tag{2.203}$$

$$E_\varphi = E_\varphi(\vartheta)\cos(\varphi) \tag{2.204}$$

or

$$E_\vartheta = -E_\vartheta(\vartheta)\cos(\varphi) \quad (2.205)$$
$$E_\varphi = E_\varphi(\vartheta)\sin(\varphi) \quad (2.206)$$

where ϑ and φ are the position coordinates on the spherical cap, it can be shown that the far-field components of the radiated field at a distance R from the origin are given by

$$E_\theta(\theta, \phi = 0) = \frac{jkR_1^2 \exp(-jkR)}{4R}\chi(\theta) \quad (2.207)$$
$$E_\phi(\theta, \phi = \pi/2) = \frac{jkR_1^2 \exp(-jkR)}{4R}\xi(\theta) \quad (2.208)$$

where

$$\chi(\theta) = \int_0^{\theta_1} [(c+a)J_0(x) + (c-a)J_2(x) + j2bJ_1(x)] \sin(\vartheta)\exp[jkR\cos(\theta)\cos(\vartheta)]\, d\vartheta \quad (2.209)$$
$$\xi(\theta) = \int_0^{\theta_1} [(e-d)J_0(x) - (e+d)J_2(x) + j2fJ_1(x)] \sin(\vartheta)\exp[jkR\cos(\theta)\cos(\vartheta)]\, d\vartheta \quad (2.210)$$
$$a = -E_\vartheta(\vartheta)\left[1 + \frac{Z_0}{Z_1}\cos(\vartheta)\cos(\theta)\right] \quad (2.211)$$
$$b = -E_\vartheta(\vartheta)\frac{Z_0}{Z_1}\sin(\vartheta)\sin(\theta) \quad (2.212)$$
$$c = -E_\varphi\left[\cos(\vartheta) + \frac{Z_0}{Z_1}\cos(\theta)\right] \quad (2.213)$$
$$d = -E_\vartheta\left[\cos(\theta) + \frac{Z_0}{Z_1}\cos(\vartheta)\right] \quad (2.214)$$
$$e = E_\varphi\left[\cos(\vartheta)\cos(\theta) + \frac{Z_0}{Z_1}\right] \quad (2.215)$$
$$f = E_\varphi \sin(\vartheta)\sin(\theta) \quad (2.216)$$
$$x = kR\sin(\vartheta)\sin(\theta) \quad (2.217)$$

and Z_1 and Z_0 are the aperture impedance and the free space impedance,respectively.

The directivity on boresight, referred to an isotropic antenna, is

$$di(\theta = 0) = \frac{kR}{2}\frac{|\chi(\theta = 0)|}{\sqrt{\zeta}} \quad (2.218)$$

where ζ is

$$\zeta = \frac{Z_0}{Z_1} \int_0^{\pi} \left[E_{\vartheta}^2(\vartheta) + E_{\varphi}^2(\vartheta) \right] \sin(\vartheta)\, d\vartheta \tag{2.219}$$

2.13 REMARKS ON ANTENNA DESIGN AND TEST

The main antenna parameters for the electric design are:

- Radiation pattern;
- Efficiency;
- Impedance;
- Bandwidth, for which the impedance and the radiation pattern are kept within given limits.

In this book we will concentrate mainly on the first two parameters.

2.13.1 Gain, Directivity, Efficiency, and Aperture Efficiency

Gain may be defined, for a given distance and direction, as the ratio between the field produced by an antenna and the field produced by an isotropic lossless antenna fed with the same power. This is the definition commonly used in surface wave antennas and is the one adopted here.

To avoid any misunderstanding, references to gains or radiation patterns expressed in logarithmic units will be denoted as dBi. For wire antennas an ideal lossless half-wavelength dipole may also be used as reference antenna although nowadays this practice appears to find fewer followers.

If $\boldsymbol{E}(\theta, \phi)$ is the free space field radiated by an aperture antenna at a distance R in the direction (θ, ϕ), and if P is power fed to the antenna, the directivity is

$$di = \frac{4\pi R^2 |\boldsymbol{E}(\theta, \phi)|^2}{Z_0 P} \tag{2.220}$$

The gain of an antenna g is the product of the directivity di by the efficiency η with which the power fed to the antenna terminals reaches the aperture:

$$g = di\, \eta \tag{2.221}$$

Efficiency is usually high (above 70%) for aperture antennas. In practical cases it may even approach unity.

A quite different concept, but nevertheless one that is often confused with antenna efficiency, is *aperture efficiency*, defined as the ratio of the directivity of an aperture and the directivity of the same aperture with a uniform, constant-phase illumination. Following this definition a uniformly illuminated aperture has a aperture efficiency of unity (or 100%). Typical values of aperture efficiency for horns are around 60% to 70%.

2.13.2 Far-Field Distance

In the application of the Kirchhoff-Huygens formulas we placed the observation point in the far field that is at a large distance from the aperture. In real life, the most common definition of far field distance is one that exceeds $d_{\text{far}} = 2d_a^2/\lambda$, where d_a is the largest aperture dimension, normal to the direction of propagation.

If we take a spherical wave at a distance d_{far} from its center it is easy to find out that the maximum phase error on a plane aperture, normal to the direction of propagation, with a maximum dimension d_a is $\pi/8$. For the usual aperture antennas where $d_a > \lambda$, the above far-field condition ensures the validity of application of the far-field Kirchhoff-Huygens formulas.

2.13.3 Polarization

Polarization may be an important feature of the radiation pattern of an antenna. In practical cases linear polarization is often specified in radio relay links and circular polarization in satellite links.

In the first case one usually specifies the minimum cross polarization discrimination, that is, the ratio between the maximum antenna gain and the maximum gain for a cross polarized field. Values of 20 to 30 dB are often required.

For circular polarization the *axial ratio*, defined as the ratio between the major and the minor axes of the polarization ellipsis, is often specified together with the direction of rotation. A minimum value of cross-polarization discrimination may also be imposed. Typical values of the axial ratio are better than 3 dB.

2.13.4 Phase Center

Sometimes it is useful to make use of an *antenna phase center* defined as a point such that the radiation patterns obtained when the antenna is turned about it exhibit constant phase at least for a considerable part of the main lobe. Typical examples of the use of the phase center are feeds for reflector antennas and lenses.

For rectangular or circular plane apertures with no phase error the phase center lies on the aperture center. If the aperture exhibits a quadratic phase error, such as is the case in horns, the phase center is usually located slightly behind the aperture. Complex antenna with multiple radiating surfaces such as the solid dielectric rectangular waveguide do not have a phase center.

The position of the phase center may be found experimentally by trial and error, which is a very lengthy and cumbersome process. There is, however, a simple, albeit little-known alternative due to Hu [9] that is worth describing.

Consider an antenna rotated about an arbitrary point O coaxially placed at a distance d_c behind the antenna phase center C, as shown in Figure 2.18. Let θ be the angle with boresight for the current point on the measured radiation pattern and let r_c and r_o be the distances from it to C and O, respectively. Since

$$r_c = \sqrt{r_o^2 + d^2 - 2r_o d\cos\theta} \tag{2.222}$$

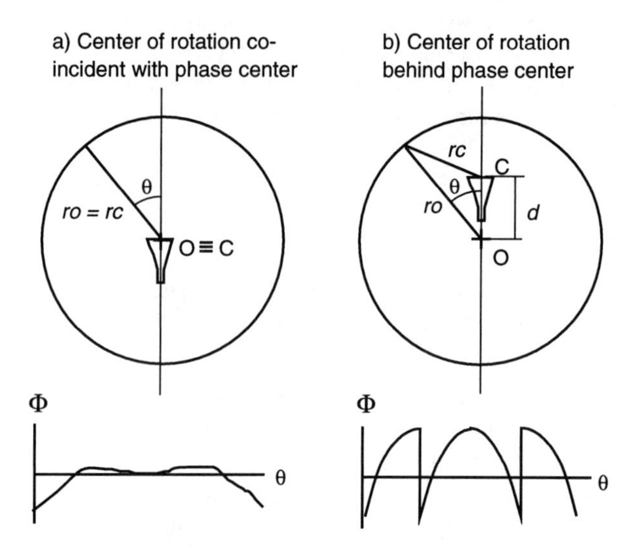

Figure 2.18: Schematic diagram to define the antenna phase center in respect to its center of rotation.

In the far field $d << r_o$ and $d << r_c$ and thus, from (2.222) we get

$$r_c \approx r_o - d\cos\theta \tag{2.223}$$

If C is the phase center, the phase Φ_o measured at each point (r_o, θ) is given by

$$\begin{aligned}\Phi_o &= kr_o \\ &= kr - kd\cos\theta\end{aligned} \tag{2.224}$$

Deriving the phase Φ_o with respect to $\cos\theta$ and noting that r_o is a constant, we get

$$d = -\frac{1}{k}\frac{\partial(\Phi_o)}{\partial(\cos\theta)} \tag{2.225}$$

We should recall that (2.224) requires the antenna to have a well-defined phase center. This may be easily confirmed from the measured phase pattern $\Phi_o(\theta)$ if we plot it as $\Phi_o(\cos\theta)$. If there is a range of θ for which the slope of $\Phi_o(\cos\theta)$ may be taken as constant then the antenna has a phase center located according to (2.225).

2.13.5 Materials

Real-life antennas make use of real conductors and dielectrics. It is thus useful to recall some of their characteristics.

A *conductor* is defined as a material with a large number of free detachable electrons or a material having high conductivity σ (ohm^{-1}m^{-1}). Typical conductors used in antennas are [8]:

- Silver ($\sigma = 6.14 \times 10^7$ ohm^{-1}m^{-1});
- Copper ($\sigma = 5.8 \times 10^7$ ohm^{-1}m^{-1});
- Aluminum ($\sigma = 3.54 \times 10^7$ ohm^{-1}m^{-1}).

A *dielectric* is a material with few free detachable electrons or with a low value of conductivity. At high frequencies a common measure of low conductivity is the loss tangent, defined as $\tan\delta = \sigma/(\omega\varepsilon)$. Examples of dielectrics used in antennas at frequencies around 10 GHz are [10]:

- Teflon ($\tan\delta \approx 4 \times 10^{-4}$, $\bar{\varepsilon} \approx 2.1$);
- Polyethylene ($\tan\delta \approx 5 \times 10^{-4}$, $\bar{\varepsilon} \approx 2.26$);
- Polystyrene ($\tan\delta \approx 6 \times 10^{-4}$, $\bar{\varepsilon} \approx 2.54$);
- Plexiglas ($\tan\delta \approx 2 \times 10^{-2}$, $\bar{\varepsilon} \approx 2.53$).

Expanded polystyerene, which may be obtained rather inexpensively as packing material, may be used in the fabrication of low-permittivity dielectric antenna prototypes for indoor use. The relative dielectric constant $\bar{\varepsilon}$ is related with the specific weight w (in kg/dm^3) by the following linear relation:

$$\bar{\varepsilon} = 1 + 1.48w \tag{2.226}$$

Taking a commonly found density value $w = 0.032$ kg/dm^3 we have $\bar{\varepsilon} = 1.047$. The loss tangent for expanded polystyerene foam dielectric is very low, on the order of 10^{-5} or less.

In many cases conductivity for a good conductor may be taken as infinity and for a good dielectric as zero.

Furthermore, throughout this book the media are assumed to be linear, homogeneous, and isotropic.

REFERENCES

[1] Harrington, R. F., (1961), *Time Harmonic Electromagnetic Fields*, McGraw-Hill, New-York.

[2] Silver S. (ed.), (1965), *Microwave Antenna Theory and Design*, Dover, New-York.

[3] Faro, M. A., (1979), *Propagação e Radiação de Ondas Electromagnéticas: I — Ondas e Meios Materiais*, Técnica, AEIST, Lisboa.

[4] Deschamps, G. A. (1972), Ray techniques in electromagnetics, Proceedings of the IEEE, Vol. 60, No. 9, pp. 1022–1035.

[5] Franz, W., (1948), Zur formulierung des Huygensschen Prinzips, *Zeitung Naturforsch. A.*, Vol. 3a, pp. 500–506.

[6] Chen-To Tai, (1972), Kirchhoff Theory: Scalar, Vector or Dyadic? *IEEE Transactions on Antennas and Propagation*, January, pp. 114–115.

[7] Salema, C., (1989), A Numerically Efficient Algorithm to Compute Radiation Patterns of Large Aperture Antennas, *Microwave and Optical Technology Letters*, Vol. 2, No. 7, pp. 260–264

[8] Chatterjee, R., (1988), *Advanced Microwave Engineering*, John Wiley & Sons, New-York.

[9] Hu, Y. Y., (1961), Determination of phase centre of radiating systems, *Journal of the Franklin Institute*, Vol. 271, pp. 21–29

[10] International Telephone & Telegraph Corporation, (1964), *Reference Data for Radio Engineers*, American Book–Stratford Press, New York.

Chapter 3

Dielectric Waveguide

3.1 INTRODUCTION

The solid rectangular dielectric waveguide, henceforth simply called the dielectric waveguide, is possibly one of the simplest dielectric antennas. This is particularly true if one thinks of the standard metallic rectangular waveguide operated in the $TE_{1,0}$ mode as the feeder (or launcher). Figure 3.1 shows a possible configuration of such an antenna where the external dielectric waveguide dimensions are the same as the internal dimensions of the metallic waveguide.

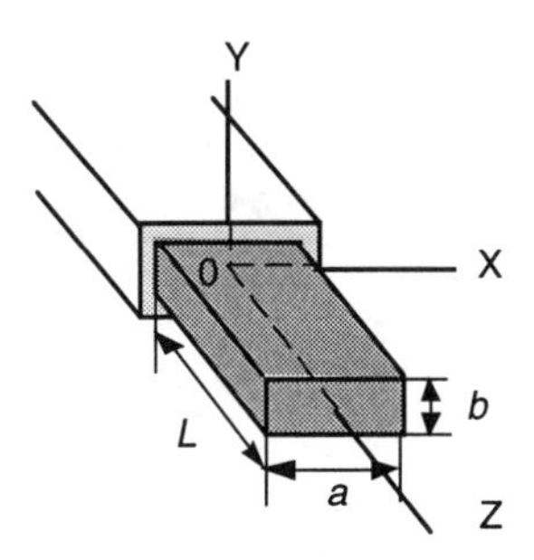

Figure 3.1: Dielectric waveguide antenna excited by a rectangular metallic waveguide.

Practical antennas will hardly make use of such an arrangement because, with some simple modifications, better performances may be achieved. However, the dielectric waveguide provides the basic information required for devising improved antennas and illustrates a number of points that may prove intractable in more complex structures.

Unlike metallic walls, boundary conditions at dielectric walls do not constrain the tangential component of the electric field to vanish and, for moderate to low values of the relative dielectric permittivity ($\bar{\varepsilon}_1 < 3$), affect normal and tangential field components in a not too unlikely way. One is thus lead to expect an improved radiation pattern symmetry for the dielectric waveguide. Additionally, since the fields in the dielectric waveguide extend beyond the dielectric boundaries, increased directivity (compared to the waveguide free end) should also be expected.

The first work on the dielectric waveguide antenna, then commonly referred to as the polyrod[1] antenna, is attributed to G. C. Southworth [1] in 1940, by Mueller [2]. A few years later in a paper describing the work done from 1941 to 1944 at the Bell Telephone Laboratories, Holmdel, New Jersey, Mueller approached the design of a dielectric waveguide antenna of uniform rectangular cross section by "establishing analogies with array theory, coupled with existing knowledge about transmission in uniform dielectric wires." Tapering down the rod (i.e., reducing its cross section toward the free end) for about half of its length not only increased the gain but also decreased the sidelobe level to about -18 dB below boresight. Boresight gains in the range of 15 to18 dBi were achieved for rod lengths between about 6 and 9 λ. Since the observed results agreed closely with the predicted ones he deemed that "a rigorous field solution has not so far appeared necessary." A similar approach was taken by Shiau [3] in the design of dielectric antennas for millimeter-wave (60 GHz) integrated circuits. Further use of tapered dielectric waveguide antennas in an integrated circuit for the frequency range between 210 and 240 GHz is reported by Yao [4].

Watson and Horton [5] put forward the first theoretical analysis of the radiation pattern of tapered dielectric rods of rectangular cross section. They applied the equivalence theorem to the whole volume of the rod assuming that the fields inside the rod would be the same as those inside the metallic waveguide used as the launcher. They justified their approach by the fact that the dielectric rod has the same dimensions of the metallic waveguide where it is inserted for a considerable distance. Experimental results are only given for the E-plane patterns and the agreement with theory is acceptable for fairly thin rods ($\approx 0.2\lambda$).

Further work on dielectric waveguide antennas had to wait for a better description of the fields in the dielectric. In 1969 Marcatili [6] presented an approximate solution of Maxwell's equations for an infinite dielectric waveguide of rectangular cross section, based on a rectangular coordinate system, which describes the fields in terms of cosine functions inside the dielectric and decaying exponentials outside. The "effective dielectric constant," put forward by McLevige [7], while keeping the simplicity of application, extends and improves Marcatili's results, particularly near the cutoff region.

In a paper [8] published simultaneously with Marcatili's, Goell took a different approach and, based on a cylindrical coordinate system, derived the fields in terms of space harmonics described by Bessel functions and modified Bessel functions multiplied by trigonometric functions (such as is presented in Chapter 5). He matched the boundary conditions at a number of specially selected points by assuming a few (five to nine) space harmonics. This ensures adequate precision for low to moderate (up to about five) aspect ratios. Since Goell's results agree well with Marcatili's for most cases of practical interest

---

[1]From polystyrene rod.

in antenna work when both methods are valid and given the computational simplicity of the latter, this is the preferred approach by most researchers.

Proceeding along somewhat similar lines, Eyges [9] extended Goell's results to dielectric waveguides of arbitrary cross sectional shapes. Again results agree quite closely with those of Marcatili for rectangular dielectric waveguides with the usual aspect ratios and frequency ranges.

Morita [10] derived integral field expressions for cylindrical dielectric waveguides of arbitrary cross section and then proceeded to apply these expressions to rectangular waveguides assuming the fields are expressed as a sum of cylindrical space harmonics and that in the resulting modes one transversal component will be dominant, an approach that in practice yields results similar to those of Marcatili for transverse electric (TE) to X and TE to Y modes.

Taking Marcatili's description for the fields inside a dielectric waveguide, Sen [11, 12] applied both the equivalence theorem and the two aperture theory to derive the radiation pattern of a dielectric waveguide antenna. Paulit [13] extended this work to the tapered case.

Kobayashi [14] presented experimental results for tapered dielectric waveguide antennas for millimeter-wave applications designed according to Zucker's [15] maximum gain design principle and provided guidelines for the design of low-sidelobe-level antennas.

Due to its simplicity coupled with sufficient accuracy, the combination of the method of Marcatili and the "effective dielectric constant" method is the basis for most analyses of the dielectric waveguide antenna, even though it leads to serious underestimation of the fields outside the dielectric for small dielectric cross sections.

In the following discussion we start by writing the field equations in terms of the magnetic and electric vector potentials. Then, applying boundary conditions, we derive characteristic equations for the main sets of modes, both following Marcatili's method and using a modified method that provides a more accurate description of the transverse field components outside the dielectric waveguide while leading to similar values inside it. After defining the cutoff frequencies we look closer into field structure and, using modal matching techniques, we evaluate the excitation efficiency for a simple launcher made up of an open-ended metallic waveguide operated in the $TE_{1,0}$ mode. We derive the radiation pattern of a dielectric waveguide antenna including the effects of the launcher and of reflections at the waveguide's free end. Finally, we introduce the shielded dielectric waveguide antenna, a simple modification of the dielectric waveguide antenna with a much-improved radiation pattern. In all cases predicted radiation patterns are compared and found to be in excellent agreement with measurements.

3.2 FIELD STRUCTURES

3.2.1 Field Equations in Terms of Vector Potentials

Consider an infinite lossless dielectric waveguide, a prism with rectangular cross section, with permittivity ε_1 and permeability μ_0 immersed in otherwise free space with constants ε_0 and μ_0, and a coaxial system of rectangular coordinates as shown in Figure 3.1. In Figure 3.2 we represent the waveguide cross section and number the regions inside and outside the dielectric waveguide.

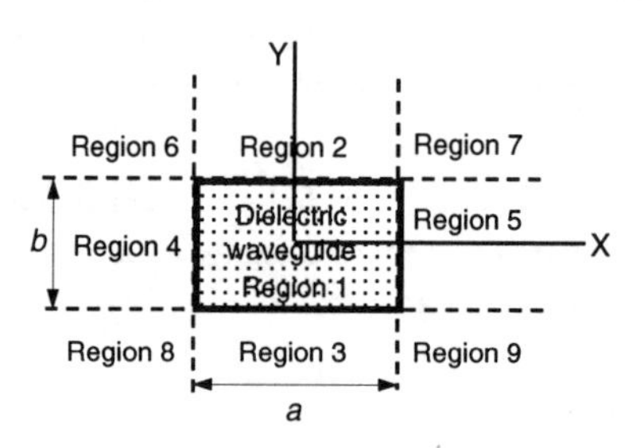

Figure 3.2: Cross section of the dielectric waveguide antenna.

Taking for the fields an $\exp(j\omega t)$ dependence it is possible (see Section 2.5) to write the electric and magnetic fields $\boldsymbol{E}$ and $\boldsymbol{H}$ either as a function of only the magnetic vector potential $\boldsymbol{A}$, as in (2.61) and (2.62); or as a function of the electric vector potential $\boldsymbol{F}$, as in (2.63) and (2.64); or still, more generally, as a sum of the previous expressions:

$$\boldsymbol{E} = -\frac{1}{\varepsilon}\nabla \times \boldsymbol{F} - j\omega \boldsymbol{A} + \frac{1}{j\omega\varepsilon\mu}\nabla(\nabla \cdot \boldsymbol{A}) \tag{3.1}$$

$$\boldsymbol{H} = \frac{1}{\mu}\nabla \times \boldsymbol{A} - j\omega \boldsymbol{F} + \frac{1}{j\omega\varepsilon\mu}\nabla(\nabla \cdot \boldsymbol{F}) \tag{3.2}$$

where the vector potentials $\boldsymbol{A}$ and $\boldsymbol{F}$ are chosen according to the excitation.

As will be shown in the following discussion, in the dielectric waveguide we may derive two basic sets of modes: one with a linearly polarized transversal electric field and the other with a linearly polarized transversal magnetic field. In each case these modes may be either symmetric (even) or antisymmetric (odd) with respect to coordinate axes X and Y.

Before we continue, it is instructive to point out why there is no simple and exact solution to the field equations in a dielectric waveguide [16]. Take, for instance (Figure 3.2), the boundary between regions 1 and 2, where $y = +b/2$ and $|x| <= a/2$, where we must have

$$\varepsilon_0 E_{y_2} = \varepsilon_1 E_{y_1} \tag{3.3}$$

On the other hand, on the boundary between regions 1 and 4

$$E_{y_4} = E_{y_1} \tag{3.4}$$

On the corner, where regions 1, 2, and 4 meet, (3.3) and (3.4) must hold simultaneously, which means that the electric field cannot be continuous at this point. This behavior makes it impossible to get a simple analytical solution to field equations. Even approximate (numeric) solutions converge slowly at this point.

Marcatili preferred continuous solutions for the electric field and thus did not impose the continuity of the normal component of the electric displacement at the dielectric boundaries. Also, he was mainly interested in the fields inside the waveguide and on its free-space boundaries (see Figure 3.2) and so did not consider regions 6 through 9 nor did he apply boundary conditions between these regions and their neighbors.

Besides breaking down when the fields in the dielectric waveguide extend far away from the dielectric, that is, when the dielectric cross section dimensions are small compared to the free-space wavelength, this approximation underestimates field intensities outside the dielectric waveguide. A better description of the main transversal field components is obtained if we impose the continuity of the normal component of the electric displacement on the dielectric boundaries at $y = \pm b/2$; we do this at the cost of the continuity of the longitudinal electric field.

3.2.2 TE to X Modes

If the dielectric waveguide is excited by a Y polarized field, such as the $TE_{1,0}$ mode of a rectangular metallic waveguide, it is natural to choose the vector potentials that lead to fields inside the dielectric waveguide with the same properties, that is, an electric field transversal to the X axis— in short TE to X— or a magnetic field transversal to the Y axis — in short TM to Y . Following Marcatili we will deal separately with each of these modes starting with the TE to X modes.

Taking

$$\boldsymbol{F} = \varepsilon \Psi \, \boldsymbol{x} \tag{3.5}$$

$$\boldsymbol{A} = 0 \tag{3.6}$$

and applying (3.5) and (3.6) into (3.1) and (3.2) in rectangular coordinates we get

$$E_x = 0 \tag{3.7}$$

$$E_y = -\frac{\partial \Psi}{\partial z} \tag{3.8}$$

$$E_z = \frac{\partial \Psi}{\partial y} \tag{3.9}$$

$$H_x = \frac{1}{j\omega\mu}\left(\frac{\partial^2 \Psi}{\partial x^2} + k^2\Psi\right) \tag{3.10}$$

$$H_y = \frac{1}{j\omega\mu}\frac{\partial^2 \Psi}{\partial x \partial y} \tag{3.11}$$

$$H_z = \frac{1}{j\omega\mu}\frac{\partial^2 \Psi}{\partial x \partial z} \tag{3.12}$$

Assuming separation of variables (which is here an approximation) and unity amplitude and omitting a common $\exp(j\omega t)$ factor for a wave propagating along the positive direction of the Z axis, inside the dielectric waveguide we have

$$\Psi_1 = \mathrm{f}(k_{x_1}x)\ \mathrm{g}(k_{y_1}y)\exp(-j\beta z) \tag{3.13}$$

where functions f and g are, in principle, any of the circular functions (cosine or sine) or the decaying exponential function, provided the resulting fields match the boundary conditions [17], β is the longitudinal propagation constant, and k_{x_1} and k_{y_1} are the transversal propagation constants inside the dielectric along the X and the Y axes respectively. Outside the dielectric, for the fields to vanish at infinity the functions f and g must be the decaying exponential function. Inside the dielectric the choice depends on the excitation. Modes are said to be symmetric (even) or antisymmetric (odd) according to the choice of the cosine or the sine function, respectively.

When the dielectric waveguide is excited by the $TE_{1,0}$ mode of a rectangular metallic waveguide, which is symmetric with respect to both X and Y axes, the cosine function should be chosen inside the dielectric

$$\Psi_1 = \cos(k_{x_1}x)\cos(k_{y_1}y)\exp(-j\beta z) \tag{3.14}$$

As shown in Section 2.7, when using rectangular coordinates the amplitude of the magnetic vector potential Ψ obeys the Helmholtz equation:

$$\nabla^2\Psi + k^2\Psi = 0 \tag{3.15}$$

where

$$\begin{aligned} k &= \omega\sqrt{\varepsilon\mu} \\ &= \frac{2\pi}{\lambda} \end{aligned} \tag{3.16}$$

and λ is the "free-space" wavelength in the region.

Imposing (3.15) on the magnetic vector potential given by (3.14) we get

$$\beta = \sqrt{k_1^2 - k_{x_1}^2 - k_{y_1}^2} \tag{3.17}$$

In the regions above and below the waveguide (regions 2 and 3), taking symmetry and the need for fields to vanish at infinity, we must write

$$\Psi_2 = \zeta \cos(k_{x_1} x) \exp(-k_{y_0} y) \exp(-j\beta z) \tag{3.18}$$
$$\Psi_3 = \zeta \cos(k_{x_1} x) \exp(+k_{y_0} y) \exp(-j\beta z) \tag{3.19}$$

which for brevity may be simply written as

$$\Psi_{2,3} = \zeta \cos(k_{x_1} x) \exp(-k_{y_0} |y|) \exp(-j\beta z) \tag{3.20}$$

Imposing again (3.15) on potentials given by (3.18) and (3.19) we get

$$\beta = \sqrt{k_0^2 - k_{x_1}^2 + k_{y_0}^2} \tag{3.21}$$

where k_{y_0} is the transversal propagation constant outside the dielectric and along the Y axis.

In the regions in front and behind the waveguide (regions 4 and 5) for regions 2 and 3 we must have

$$\Psi_{4,5} = \xi \exp(-k_{x_0} |x|) \cos(k_{y_1} y) \exp(-j\beta z) \tag{3.22}$$

where now k_{x_0} is the transversal propagation constant outside the dielectric and along the X axis.

Applying again the Helmoltz equation (3.15) to the potentials in (3.22) we get

$$\beta = \sqrt{k_0^2 + k_{x_0}^2 - k_{y_1}^2} \tag{3.23}$$

The fields in each region are obtained substituting the appropriate value of Ψ in (3.7) to (3.12) where k, ε, and μ are the values for the region. Combining into a common expression the expressions for regions 2 and 3, on the one hand, and for regions 4 and 5, on the other, and dropping a common factor $\exp(-j\beta z)$ we have

$$E_{x_1} = 0 \tag{3.24}$$

$$E_{x2,3} = 0 \tag{3.25}$$

$$E_{x4,5} = 0 \tag{3.26}$$

$$E_{y_1} = +j\beta \cos(k_{x_1}x)\cos(k_{y_1}y) \tag{3.27}$$

$$E_{y2,3} = +j\zeta\beta \cos(k_{x_1}x)\exp(-k_{y0}|y|) \tag{3.28}$$

$$E_{y4,5} = +j\xi\beta \exp(-k_{x0}|x|)\cos(k_{y_1}y) \tag{3.29}$$

$$E_{z1} = -k_{y_1}\cos(k_{x_1}x)\sin(k_{y_1}y) \tag{3.30}$$

$$E_{z2,3} = \mp\zeta k_{y0}\cos(k_{x_1}x)\exp(-k_{y0}|y|) \tag{3.31}$$

$$E_{z4,5} = -\xi k_{y_1}\exp(-k_{x0}|x|)\sin(k_{y_1}y) \tag{3.32}$$

$$H_{x_1} = \frac{k_1^2 - k_{x1}^2}{j\omega\mu_0}\cos(k_{x_1}x)\cos(k_{y_1}y) \tag{3.33}$$

$$H_{x2,3} = \zeta\frac{k_0^2 - k_{x1}^2}{j\omega\mu_0}\cos(k_{x_1}x)\exp(-k_{y0}|y|) \tag{3.34}$$

$$H_{x4,5} = \xi\frac{k_0^2 + k_{x0}^2}{\omega\mu_0}\exp(-k_{x0}|x|)\cos(k_{y_1}y) \tag{3.35}$$

$$H_{y_1} = +\frac{k_{x_1}k_{y_1}}{j\omega\mu_0}\sin(k_{x_1}x)\sin(k_{y_1}y) \tag{3.36}$$

$$H_{y2,3} = \pm\zeta\frac{k_{x_1}k_{y0}}{j\omega\mu_0}\sin(k_{x_1}x)\exp(-k_{y0}|y|) \tag{3.37}$$

$$H_{y4,5} = \pm\xi\frac{k_{x0}k_{y1}}{j\omega\mu_0}\exp(-k_{x0}|x|)\sin(k_{y_1}y) \tag{3.38}$$

$$H_{z_1} = +\frac{\beta k_{x1}}{\omega\mu_0}\sin(k_{x_1}x)\cos(k_{y_1}y) \tag{3.39}$$

$$H_{z2,3} = +\zeta\frac{\beta k_{x_1}}{\omega\mu_0}\sin(k_{x_1}x)\exp(-k_{y0}|y|) \tag{3.40}$$

$$H_{z4,5} = \pm\xi\frac{\beta k_{x0}}{\omega\mu_0}\exp(-k_{x0}|x|)\cos(k_{y_1}y) \tag{3.41}$$

In the above where the $\pm$ or $\mp$ is used, the upper sign is meant for $x, y \geq 0$ and the lower sign for $x, y < 0$.

On the dielectric boundaries we must have the continuity of the tangential components of the electric and magnetic fields and the continuity of the normal component of the electric displacement and magnetic flux density.

Let us start with the boundary between regions 1 and 2 and 1 and 3, that is, $y = \pm b/2$ and $|x| <= a/2$. For Ez, from (3.30) and (3.31) we conclude that

$$\zeta = \frac{k_{y_1}}{k_{y_0}} \frac{\sin(k_{y_1} b/2)}{\exp(-k_{y_0} b/2)} \tag{3.42}$$

Similarly, for the tangential components of the magnetic field, Hx and Hz, from (3.33) and (3.34) and from (3.39) and (3.40), respectively, we find

$$\zeta = \frac{k_1^2 - k_{x_1}^2}{k_0^2 - k_{x_1}^2} \frac{\cos(k_{y_1} b/2)}{\exp(-k_{y_0} b/2)} \tag{3.43}$$

$$\zeta = \frac{\cos(k_{y_1} b/2)}{\exp(-k_{y_0} b/2)} \tag{3.44}$$

For the normal component of the magnetic field, Hy, from (3.36) and (3.37) we again get (3.42).

Finally, for the normal component of the electric field, Ey, from (3.27) and (3.28), we get

$$\zeta = \frac{\varepsilon_1}{\varepsilon_0} \frac{\cos(k_{y_1} b/2)}{\exp(-k_{y_0} b/2)} \tag{3.45}$$

It is obvious that (3.43) through (3.45) cannot all simultaneously be met. If, following [18] we choose to match Ez and Hz, coupling the value of ζ given in (3.42) and (3.44) we get

$$\tan\left(k_{y_1} \frac{b}{2}\right) = \frac{k_{y_0}}{k_{y_1}} \tag{3.46}$$

Substituting in (3.46) the value of k_{y_0} taken from (3.21) where β is given in (3.17) we finally get an equation, also called characteristic or modal equation, on k_{y_1}:

$$k_{y_1} \tan\left(k_{y_1} \frac{b}{2}\right) = \sqrt{k_1^2 - k_0^2 - k_{y_1}^2} \tag{3.47}$$

For the boundary between regions 1 and 4 and 1 and 5 we get, for E_y and E_z

$$\xi = \frac{\cos(k_{x_1} a/2)}{\exp(-k_{x_0} a/2)} \tag{3.48}$$

and for H_y and H_z

$$\xi = \frac{k_{x_1}}{k_{x_0}} \frac{\sin(k_{x_1}a/2)}{\exp(-k_{x_0}a/2)} \tag{3.49}$$

and, finally, for H_x, using (3.17) and (3.23)

$$\begin{aligned} \xi &= \frac{k_1^2 - k_{x_1}^2}{k_0^2 + k_{x_0}^2} \frac{\cos(k_{x_1}a/2)}{\exp(-k_{x_0}a/2)} \\ &= \frac{\cos(k_{x_1}a/2)}{\exp(-k_{x_0}a/2)} \end{aligned} \tag{3.50}$$

Now we may simultaneously match all boundary conditions and get

$$k_{x_1} \tan\left(k_{x_1}\frac{a}{2}\right) = k_{x_0} \tag{3.51}$$

Substituting in (3.51) the value of k_{x_0} from (3.23), where β is given in (3.17), we get a characteristic equation on k_{x_1}:

$$k_{x_1} \tan\left(k_{x_1}\frac{a}{2}\right) = \sqrt{k_1^2 - k_0^2 - k_{x_1}^2} \tag{3.52}$$

From the preceding discussion it is now obvious that the TE to X mode, as described, is only an approximate solution to Maxwell's equations in the dielectric waveguide.

Let us now check that the choice of electric and magnetic vector potentials (3.5) and (3.6) was indeed appropriate to the field distribution exciting the dielectric waveguide. For the $TE_{1,0}$ of the metallic waveguide in free space we know that, dropping a common $\exp(j\omega t)$ factor, and taking unity amplitude

$$E_x = 0 \tag{3.53}$$

$$E_y = \cos(k_x x)\exp(-j\beta z) \tag{3.54}$$

$$E_z = 0 \tag{3.55}$$

$$H_x = -\frac{\beta}{k_0 Z_0}\cos(k_x x)\exp(-j\beta z) \tag{3.56}$$

$$H_y = 0 \tag{3.57}$$

$$H_z = -j\frac{k_0}{kx Z_0}\left[1 - \left(\frac{\beta}{k_0}\right)^2\right]\sin(k_x x)\exp(-j\beta z) \tag{3.58}$$

where Z_0 is the free-space wave impedance:

$$Z_0 = \sqrt{\frac{\mu_0}{\varepsilon_0}} \tag{3.59}$$

k_x the transversal wave constant:

$$k_x = \frac{\pi}{a} \tag{3.60}$$

and β the axial propagation constant:

$$\beta = k_0 \sqrt{1 - \left(\frac{k_x}{k_0}\right)^2} \tag{3.61}$$

Pairwise inspection of (3.24), (3.27), (3.30), (3.33), (3.36), and (3.39) with (3.53) to (3.58) shows that the copolar components of the electric field (E_y) and the magnetic field (H_x) exhibit a similar variation with respect to the symmetry about the X and the Y axes. The cross-polar component of the electric field (E_x) is null in both cases, whereas for the magnetic field the cross-polar component (H_y) is null in one case and antisymmetric about the origin in the other case. We may thus conclude that the choice of vector potentials was correct.

3.2.3 TM to Y Modes

As was noted before (Section 3.2.2) in the $TE_{1,0}$ mode of a rectangular metallic waveguide we have both $E_x = 0$ and $H_y = 0$, that is, the field is both TE to X and TM to Y. In the previous section we choose the vector potentials to yield TE to X modes in the dielectric. Now we will take a different choice of vector potentials:

$$\boldsymbol{F} = 0 \tag{3.62}$$

$$\boldsymbol{A} = \mu\Psi \, \boldsymbol{y} \tag{3.63}$$

which yields a field that is transversal magnetic to Y (TM to Y).

Proceeding as for the TE to X modes we get the following expressions for the fields in terms of Ψ:

$$E_x = \frac{1}{j\omega\varepsilon}\frac{\partial^2\Psi}{\partial x \partial y} \tag{3.64}$$

$$E_y = \frac{1}{j\omega\varepsilon}\left(\frac{\partial^2\Psi}{\partial y^2} + k^2\Psi\right) \tag{3.65}$$

$$E_z = \frac{1}{j\omega\varepsilon}\frac{\partial^2\Psi}{\partial y \partial z} \tag{3.66}$$

$$H_x = -\frac{\partial\Psi}{\partial z} \tag{3.67}$$

$$H_y = 0 \tag{3.68}$$

$$H_z = \frac{\partial \Psi}{\partial x} \tag{3.69}$$

Taking as before:

$$\Psi_1 = \cos(k_{x_1} x)\cos(k_{y_1} y)\exp(-j\beta z) \tag{3.70}$$

$$\Psi_{2,3} = \zeta \cos(k_{x_1} x)\exp(-k_{y_0}|y|)\exp(-j\beta z) \tag{3.71}$$

$$\Psi_{4,5} = \xi \exp(-k_{x_0}|x|)\cos(k_{y_1} y)\exp(-j\beta z) \tag{3.72}$$

which, dropping a common factor $\exp(-j\beta z)$, lead to the following fields:

$$E_{x_1} = \frac{k_{x_1} k_{y_1}}{j\omega\varepsilon_1}\sin(k_{x_1} x)\sin(k_{y_1} y) \tag{3.73}$$

$$E_{x2,3} = \pm\zeta \frac{k_{x_1} k_{y0}}{j\omega\varepsilon_0}\sin(k_{x_1} x)\exp(-k_{y_0}|y|) \tag{3.74}$$

$$E_{x4,5} = \pm\xi \frac{k_{x0} k_{y_1}}{\omega\varepsilon_0}\exp(-k_{x_0}|x|)\sin(k_{y_1} y) \tag{3.75}$$

$$E_{y_1} = \frac{\left(k_1^2 - k_{y_1}^2\right)}{j\omega\varepsilon_1}\cos(k_{x_1} x)\cos(k_{y_1} y) \tag{3.76}$$

$$E_{y2,3} = \zeta \frac{\left(k_0^2 + k_{y0}^2\right)}{j\omega\varepsilon_0}\cos(k_{x_1} x)\exp(-k_{y_0}|y|) \tag{3.77}$$

$$E_{y4,5} = \xi \frac{\left(k_0^2 - k_{y_1}^2\right)}{j\omega\varepsilon_0}\exp(-k_{x_0}|x|)\cos(k_{y_1} y) \tag{3.78}$$

$$E_{z_1} = +\frac{\beta k_{y_1}}{\omega\varepsilon_1}\cos(k_{x_1} x)\sin(k_{y_1} y) \tag{3.79}$$

$$E_{z2,3} = \pm\zeta \frac{\beta k_{y_2}}{\omega\varepsilon_0}\cos(k_{x_1} x)\exp(-k_{y_0}|y|) \tag{3.80}$$

$$E_{z4,5} = +\xi \frac{\beta k_{y_1}}{\omega\varepsilon_0}\exp(-k_{x_0}|x|)\sin(k_{y_1} y) \tag{3.81}$$

$$H_{x_1} = j\beta\cos(k_{x_1} x)\cos(k_{y_1} y) \tag{3.82}$$

$$H_{x2,3} = j\zeta\beta\cos(k_{x_1} x)\exp(-k_{y_0}|y|) \tag{3.83}$$

$$H_{x4,5} = j\xi\beta\exp(-k_{x_0}|x|)\cos(k_{y_1} y) \tag{3.84}$$

$$H_{y_1} = 0 \tag{3.85}$$

$$H_{y2,3} = 0 \tag{3.86}$$

$$H_{y4,5} = 0 \tag{3.87}$$

$$H_{z_1} = -k_{x_1}\sin(k_{x_1} x)\cos(k_{y_1} y) \tag{3.88}$$

$$H_{z2,3} = -\zeta k_{x_1}\sin(k_{x_1} x)\exp(-k_{y_0}|y|) \tag{3.89}$$

$$H_{z4,5} = \mp\xi k_{x0}\exp(-k_{x_0}|x|)\cos(k_{y_1} y) \tag{3.90}$$

Imposing the Helmoltz equation (3.15) we find again the same equations relating β, k_{x_1}, k_{x_0}, k_{y_1}, and k_{y_0} that we had before, that is, (3.17), (3.21), and (3.23).

Applying the boundary conditions and matching all the field components on the boundaries between regions 1 and 2 and 1 and 3 and the normal component and one tangential component (along the Z axis) of the electric field and the tangential components of the magnetic field on the boundary between regions 1 and 4 and 1 and 5 we get the following characteristic equations for the TM to Y modes:

$$\frac{\varepsilon_1}{\varepsilon_0} k_{x_1} \tan\left(k_{x_1}\frac{a}{2}\right) = \sqrt{k_1^2 - k_0^2 - k_{x_1}^2} \tag{3.91}$$

$$k_{y_1} \tan\left(k_{y_1}\frac{b}{2}\right) = \frac{\varepsilon_1}{\varepsilon_0}\sqrt{k_1^2 - k_0^2 - k_{y_1}^2} \tag{3.92}$$

3.2.4 mTE to X and mTM to Y Modes

Let us take again the TE to X mode fields (see Section 3.2.2). Applying the boundary conditions between regions 1 and 2 and 1 and 3 if we choose to match the normal components of the electric displacement (εE_y) and of the magnetic field (H_y) and the longitudinal component of the electric field (E_z) we get the following modified characteristic equation on k_{y_1}:

$$k_{y_1} \tan\left(k_{y_1}\frac{b}{2}\right) = \frac{\varepsilon_1}{\varepsilon_0}\sqrt{k_1^2 - k_0^2 - k_{y_1}^2} \tag{3.93}$$

On the boundaries between regions 1 and 4 and 1 and 5 all the boundary conditions were met and so we will keep the same characteristic equation (3.52), which for convenience is repeated here:

$$k_{x_1} \tan\left(k_{x_1}\frac{a}{2}\right) = \sqrt{k_1^2 - k_0^2 - k_{x_0}^2} \tag{3.94}$$

To distinguish between the TE to X fields where the propagation constants are solutions to (3.52) and (3.47) from those which result from (3.95) and (3.47) the latter will be called modified transversal electric to X, in short mTE to X.

If we now take the TM to Y mode fields (see Section 3.2.3) and when applying the boundary conditions between regions 1 and 4 and 1 and 5 decide to match the normal component of the electric displacement (εE_x) and the normal (H_x) and the longitudinal components (H_z) of the magnetic field we get a modified characteristic equation on k_{x_1}:

$$k_{x_1} \tan\left(k_{x_1}\frac{a}{2}\right) = \sqrt{k_1^2 - k_0^2 - k_{x_0}^2} \tag{3.95}$$

At the boundaries between regions 1 and 2 and 1 and 3 all the boundary conditions were met and so the characteristic equation (3.92) will be the same:

$$k_{y_1} \tan\left(k_{y_1}\frac{b}{2}\right) = \frac{\varepsilon_1}{\varepsilon_0}\sqrt{k_1^2 - k_0^2 - k_{y_1}^2} \tag{3.96}$$

To mark the difference, the TM to Y fields where propagation constants are solutions to (3.95) and (3.92) will be named modified transversal magnetic to Y or simply mTM to Y.

It is interesting to remark that both the mTE to X and the mTM to Y modes share the same set of characteristic equations, as well as the same (functional) type of variation with x and y for the nonnull field components. As will be shown below, this modified approach leads to a much better description of the fields outside the dielectric, especially for small dielectric cross sections.

We should note that the modified characteristic equation on k_{y_1} is the same as the characteristic equation for a dielectric slab of thickness b and permittivity ε_1 when the transversal electric field is normal to the dielectric boundaries, a case for which the exact solution of Maxwell's equations is well known. Also the modified characteristic equation on k_{x_1} is the same as for a dielectric slab of thickness a and permittivity ε_1 when the transversal magnetic field is normal to the dielectric boundaries.

These similarities suggest that, as far as the propagation constants β, k_{x_1}, k_{yx_1}, k_{x_0}, and k_{y_0} are concerned, the dielectric waveguide may be viewed as a set of dielectric slabs, of thickness b and a, supporting the TE and the TM modes, respectively. A likely approach is also taken by the "effective dielectric constant" method and will be dealt with in Section 3.3.

3.2.5 Antisymmetric or Odd Modes

So far we have derived TE to X and TM to Y modes which are symmetric both in relation to the X and the Y axes, a convenient feature when the launcher is a metallic waveguide operated in the $TE_{1,0}$ mode. It is also possible to derive TE to X and TM to Y modes that are antisymmetric with respect to one or both of these axes. In this case the sine function should replace the cosine function in Ψ inside the dielectric. The derivation of the fields and characteristic equation is straightforward and is left to the reader.

3.3 TE TO X AND TM TO Y CUTOFF FREQUENCIES

Let us return to the modes more likely to be excited by a rectangular metallic waveguide operated in the $TE_{1,0}$ mode. Take, for instance, the TE to X mode whose characteristic equations we recall here:

$$k_{x_1} \tan\left(k_{x_1}\frac{a}{2}\right) = \sqrt{k_1^2 - k_0^2 - k_{x_1}^2}$$

$$k_{y_1} \tan\left(k_{y_1}\frac{b}{2}\right) = \sqrt{k_1^2 - k_0^2 - k_{y_1}^2}$$

Although we could proceed with the characteristic equations in the above form it is far more convenient to normalize all lengths by multiplying them by k_0 and all propagation constants by dividing them by k_0. We will also normalize the dielectric permittivity of the waveguide by dividing it by the free-space dielectric permittivity. Normalized values will be denoted by a bar over the symbol of the nonnormalized value.

After normalization the characteristic equations become

$$\bar{k}_{x_1} \tan\left(\bar{k}_{x_1}\frac{\bar{a}}{2}\right) = \sqrt{\bar{\varepsilon}_1 - 1 - \bar{k}_{x_1}^2} \tag{3.97}$$

$$\bar{k}_{y_1} \tan\left(\bar{k}_{y_1}\frac{\bar{b}}{2}\right) = \sqrt{\bar{\varepsilon}_1 - 1 - \bar{k}_{y_1}^2} \tag{3.98}$$

Following [17] we will provide a graphic interpretation of the characteristic equations. For a start, recalling (3.51), (3.46), (3.17), (3.21), and (3.23) we will rewrite these as

$$\bar{k}_{x_1} \tan\left(\bar{k}_{x_1}\frac{\bar{a}}{2}\right) = \bar{k}_{x_0} \tag{3.99}$$

$$\bar{\varepsilon}_1 - \bar{k}_{x_1}^2 = 1 + \bar{k}_{x_0}^2 \tag{3.100}$$

$$\tan\left(\bar{k}_{y_1}\frac{\bar{b}}{2}\right) = \bar{k}_{y_0} \tag{3.101}$$

$$\bar{\varepsilon}_1 - \bar{k}_{y_1}^2 = 1 + \bar{k}_{y_0}^2 \tag{3.102}$$

The solution of (3.97) corresponds to the values of $\bar{k}_{x_1}$ and $\bar{k}_{x_0}$ at the intersection of (3.99) with (3.100) as shown in Figure 3.3. We note that equation (3.100) represents a circumference of radius $\sqrt{\bar{\varepsilon}_1 - 1}$. A similar representation would yield the values of $\bar{k}_{y_1}$ and $\bar{k}_{y_0}$.

Since fields must decay outside the dielectric region, we restrict our representation to the positive values of the propagation constants.

From Figure 3.3 we see that (3.97) always has at least one solution, here denoted as $TE^X_{1,1}$, and that the number of solutions increases as $\bar{a}$ increases. The same applies to (3.98) and $\bar{b}$. Although Figure 3.3 refers to the TE to X modes, similar results may easily be obtained for the TM to Y mode and for the modified modes.

For the dielectric slab there is only one characteristic equation. In the present approximate formulation we have two equations, apparently independent of each other but in fact

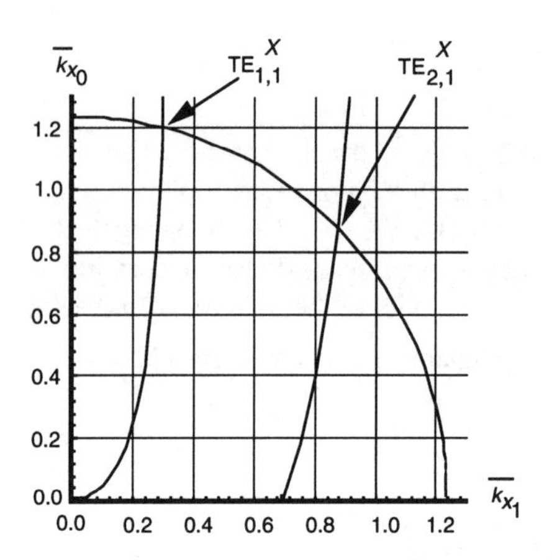

Figure 3.3: Graphic solution of the characteristic equation of the TE to X modes of a dielectric waveguide with $\bar{\varepsilon}_1 = 2.53$ and $\bar{a} = 9$ showing the first two roots.

coupled through $\bar{\beta} = \beta / k_0$:

$$\begin{aligned} \bar{\beta} &= \sqrt{\bar{k}_1^2 - \bar{k}_{x_1}^2 - \bar{k}_{y_1}^2} \\ &= \sqrt{\bar{\varepsilon}_1 - \bar{k}_{x_1}^2 - \bar{k}_{y_1}^2} \end{aligned} \quad (3.103)$$

Characteristic equations (3.97) and (3.98) may be seen as corresponding to two independent dielectric slabs, one indefinite along the X axis and the other along the Y axis.

The cutoff region corresponds to the situation where the fields do not decay outside the dielectric waveguide along the X and the Y axes, that is, when $k_{x_0} \leq 0$, and $k_{y_0} \leq 0$. Thus, to ensure propagation, both following conditions must be met:

$$\tan\left(\bar{k}_{x_1}\frac{\bar{a}}{2}\right) \geq 0 \quad (3.104)$$

$$\tan\left(\bar{k}_{y_1}\frac{\bar{b}}{2}\right) \geq 0 \quad (3.105)$$

As a consequence of having independent characteristic equations along the X and the Y axes, k_{x_1} and k_{y_1} are, in general, not simultaneously zero. Thus the value of $\bar{\beta}$ near cutoff presents a nonnegligible error.

In Figure 3.4 we represent the normalized axial propagation constant $\bar{\beta}$ versus $\bar{a}$ for a dielectric waveguide with $\varepsilon_1 = 2.53$ and aspect ratio $b/a = 4/9$ operated in the $TE^X_{1,1}$, $TM^Y_{1,1}$, and $mTE^X_{1,1}$ modes. The fact, which is not represented in the figure, that the values of $\bar{\beta}$ extend below 1 (which is physically impossible) reveals the above mentioned shortcomings of the analysis.

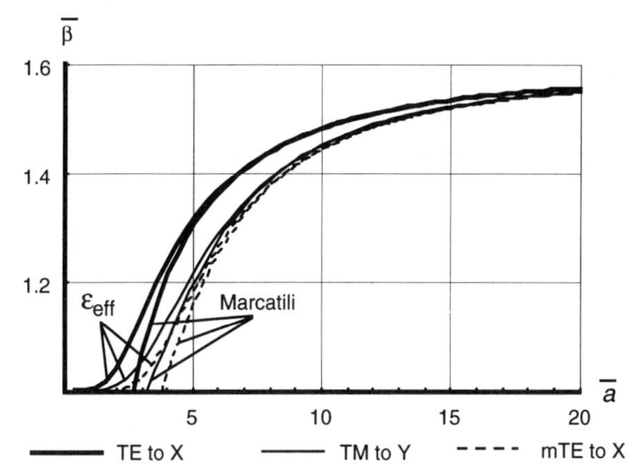

Figure 3.4: Normalized axial propagation constant for a dielectric waveguide with $\bar{\varepsilon}_1 = 2.53$ and aspect ratio $b/a = 4/9$ versus normalized waveguide width $\bar{a}$ for the $TE^X_{1,1}$, $TM^Y_{1,1}$, and $mTE^X_{1,1}$ modes. Represented values include both Marcatili's and the "effective dielectric constant" method (the latter denoted by ε_{eff}.)

A simple way of overcoming the aforementioned limitation is to use the so called "effective dielectric constant" method [7] which will be described below. This method yields results in good agreement with measurements.

Assume that in the dielectric waveguide $a > b$. In that case we start by solving the characteristic equation of a dielectric slab of thickness b, that is, (3.98). Then we define an effective dielectric constant $\bar{\varepsilon}_{eff}$ as

$$\bar{\varepsilon}_{eff} = \bar{\varepsilon}_1 - \bar{k}_{y_1}^{2} \tag{3.106}$$

Note that $\bar{\varepsilon}_{eff}$ corresponds to the value of $\bar{\beta}$ for an indefinite dielectric slab of thickness b.

In the next step we solve the characteristic equation for another dielectric slab of thickness a and permittivity $\bar{\varepsilon}_{eff}$ instead of $\bar{\varepsilon}_1$:

$$\bar{k}_{x_1} \tan\left(\bar{k}_{x_1}\frac{\bar{a}}{2}\right) = \sqrt{\bar{\varepsilon}_{eff} - 1 - \bar{k}_{x_1}^2} \tag{3.107}$$

If $a < b$ then we should start by solving the characteristic equation of a slab of thickness a, define the effective dielectric constant, and finally solve the characteristic equation for a dielectric slab of thickness b, using the effective dielectric constant just defined.

Unlike the TE to X (the TM to Y or the mTE to X) characteristic equations, now the two equations are explicitly coupled in such a way that k_{x_1} and k_{y_1} exhibit the correct behavior when approaching cutoff. This is apparent in Figure 3.4 where $\bar{\beta}$ now tends to 1 when $\bar{a}$ goes to zero. It should be noted that the discrepancy between the two methods is only significant near cutoff.

The interrelations among $\bar{k}_{x_1}$, $\bar{k}_{y_1}$, and $\bar{\beta}$ are highlighted in Figure 3.5. Here we represent, for a given aspect ratio b/a and permittivity $\bar{\varepsilon}_1$, the loci of the normalized propagation constants $\bar{k}_{x_1}$ and $\bar{k}_{y_1}$ for varying $\bar{a}$. From (3.103) we may write

$$\bar{\beta}^2 = \bar{\varepsilon}_1 - \bar{k}_{x_1}^2 - \bar{k}_{y_1}^2 \tag{3.108}$$

$$= \bar{\varepsilon}_1 - \bar{h}^2 \tag{3.109}$$

where

$$\bar{h}^2 = \bar{k}_{x_1}^2 + \bar{k}_{y_1}^2 \tag{3.110}$$

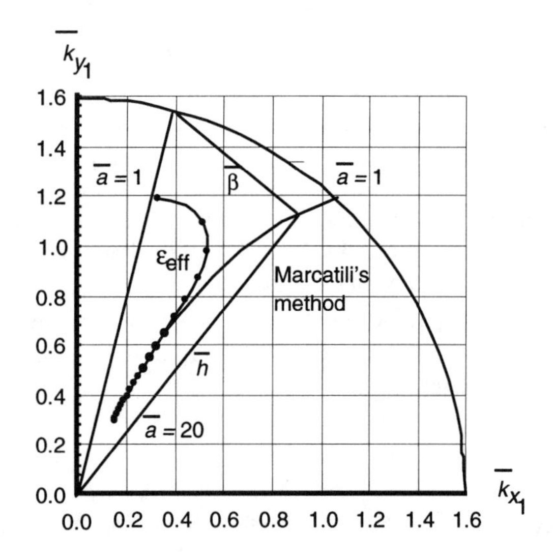

Figure 3.5: Loci of the normalized propagation constants $\bar{k}_{x_1}$ and $\bar{k}_{y_1}$ for varying $\bar{a}$ for the $TE_{1,1}^X$ mode using Marcatili's method and the "effective dielectric constant" method.

The interpretation of Figure 3.5 is straightforward if we note that (3.110) and (3.109) express relations between the sides of rectangular triangles. For each value of $\bar{a}$ we have a pair of values $\bar{k}_{x_1}$, $\bar{k}_{y_1}$ for the transverse propagation constants. The distance between the loci of the propagation constants and the circumference with center on the origin and radius $\sqrt{\bar{\varepsilon}_1}$ measured normally to $\bar{h}$ represents $\bar{\beta}$.

It should now be clear that Marcatili's method may lead to nonphysically possible values of $\bar{\beta}$ that are less than unity, a situation that does not occur when using the "effective dielectric constant" method.

3.4 APPROXIMATE SOLUTION OF CHARACTERISTIC EQUATIONS

Characteristic equations may be solved by numeric methods. However, when a solution is found there is no quick and simple method to identify the corresponding mode. We may either plot the field intensity along the X and Y axes and count the number of times the plot passes through zero or we may list all roots by ascending order and determine the position of the root we are considering. Since for antenna work we will be mostly interested in the $TE^X_{1,1}$, $TM^Y_{1,1}$, $mTE^X_{1,1}$, or $mTM^Y_{1,1}$ modes it is useful to devise a simple, even if approximate, method to solve the characteristic equations for the first mode. If more accurate solutions are required the values provided by this method may be used as a starting point in a Newton-Raphson iteration procedure.

Let us recall the characteristic equations for the TE to X mode of a dielectric waveguide, that is (3.97) and (3.98). For the $TE^X_{1,1}$ mode we may state that

$$\bar{k}_{x_1}\frac{\bar{a}}{2} < \frac{\pi}{2}$$
$$\bar{k}_{y_1}\frac{\bar{b}}{2} < \frac{\pi}{2}$$

For large $\bar{a}$ the value of $\bar{k}_{x_1}\bar{a}$ will be close to $\pi/2$ so we may state

$$\bar{k}_{x_1}\frac{\bar{a}}{2} = \frac{\pi}{2} - \delta \tag{3.111}$$

with $\delta << 1$.

Recalling that

$$\tan(x)_{x=\frac{\pi}{2}-\delta} = \frac{1}{\delta} \tag{3.112}$$

and substituting the values given in (3.111) and (3.112) in (3.97) we get

$$\left(\frac{\pi}{\bar{a}} - \frac{2\delta}{\bar{a}}\right)\frac{1}{\delta} = \sqrt{\bar{\varepsilon}_1 - 1 - \left(\frac{\pi}{a} - \frac{2\delta}{\bar{a}}\right)^2} \tag{3.113}$$

which — neglecting the expression in brackets under the square root — yields

$$\delta = \frac{\pi}{2 + \bar{a}\sqrt{\bar{\varepsilon}_1 - 1}} \tag{3.114}$$

Introducing the value of δ given in (3.114) in (3.111) we get

$$\bar{k}_{x_1} = \frac{\pi}{\bar{a}}\left(1 - \frac{1}{1 + \frac{\bar{a}}{2}\sqrt{\bar{\varepsilon}_1 - 1}}\right) \tag{3.115}$$

Similarly,

$$\bar{k}_{y_1} = \frac{\pi}{\bar{b}}\left(1 - \frac{1}{1 + \frac{\bar{b}}{2}\sqrt{\bar{\varepsilon}_1 - 1}}\right) \tag{3.116}$$

The approximation improves with increasing values of the permittivity and aperture size. For usual permittivities ($\bar{\varepsilon}_1 \approx 2.5$) the relative error is less than about 10^{-4} for normalized aperture sizes equal to or larger than about 10 and increases to about 10^{-2} for a normalized aperture size of 4.

This method is also applicable to other characteristic equations. For instance, for the $\mathrm{mTE}^X_{1,1}$ or the $\mathrm{mTM}^Y_{1,1}$ modes we get the following approximate propagation constants for the first mode:

$$\bar{k}_{x_1} = \frac{\pi}{\bar{a}}\left(1 - \frac{1}{1 + \frac{\bar{a}}{2}\sqrt{\bar{\varepsilon}_1 - 1}}\right) \tag{3.117}$$

$$\bar{k}_{y_1} = \frac{\pi}{\bar{b}}\left(1 - \frac{1}{1 + \frac{\bar{b}\bar{\varepsilon}_1}{2}\sqrt{\bar{\varepsilon}_1 - 1}}\right) \tag{3.118}$$

3.5 $\mathrm{TE}^X_{1,1}$, $\mathrm{mTE}^X_{1,1}$, $\mathrm{TM}^Y_{1,1}$, AND $\mathrm{mTM}^Y_{1,1}$ FIELDS

When the dielectric waveguide is excited by the $\mathrm{TE}_{1,0}$ field of a metallic rectangular waveguide, the first modes excited are the $\mathrm{TE}^X_{1,1}$, $\mathrm{TM}^Y_{1,1}$, $\mathrm{mTE}^X_{1,1}$, and $\mathrm{mTM}^Y_{1,1}$. These modes are the lowest order modes for which the main components of the transversal fields, E_y and H_x, have the same type of variation (cosine) along the X axis as the exciting fields. $\mathrm{TE}^X_{1,0}$ and $\mathrm{TM}^Y_{1,0}$ modes of the dielectric waveguide cannot exist because the fields would not decay outside the dielectric.

As we have shown, all the field structures of TE to X, TM to Y, mTE to X, and mTM to Y modes are similar, that is, all have the same type of functional variation both inside and outside the dielectric. For the same waveguide and the same frequency, the differences are due to the propagation constants, which are different because they are solutions to different characteristic equations. In addition, their boundary behavior is different.

In Figure 3.6 we plot E_y versus y along the Y axis, for the $\mathrm{TE}^X_{1,1}$, the $\mathrm{TM}^Y_{1,1}$, the $\mathrm{mTE}^X_{1,1}$, and the $\mathrm{mTM}^Y_{1,1}$ mode of a dielectric waveguide with $\bar{\varepsilon_1} = 2.53$, $\bar{a} = 4.49$, and $\bar{b} = 1.99$. Figure 3.6 clearly shows the incorrect boundary behavior of E_y for the $\mathrm{TE}^X_{1,1}$ mode which does not occur for the $\mathrm{TM}^Y_{1,1}$, the $\mathrm{mTE}^Y_{1,1}$, or the $\mathrm{mTM}^Y_{1,1}$ mode.

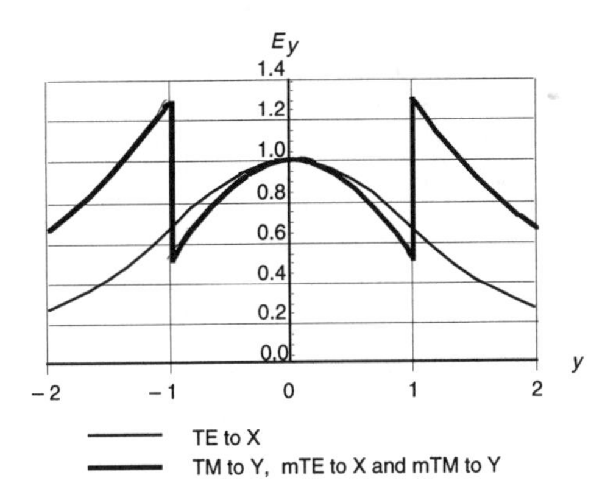

Figure 3.6: $E_y(y)$ along the Y axis for a dielectric waveguide with $\bar{\varepsilon_1} = 2.53$, $\bar{a} = 4.49$, and $\bar{b} = 1.99$.

A plot of H_y as a function of y, for $x = 0$, is shown in Figure 3.7 for the same dielectric waveguide. Here again the incorrect boundary behavior of the $\mathrm{TE}^X_{1,1}$ solution stands out. The $\mathrm{TM}^Y_{1,1}$ field has the expected continuity at the boundary while the $\mathrm{mTE}^X_{1,1}$ and the $\mathrm{mTM}^Y_{1,1}$ discontinuous are much less severe than for the $\mathrm{TE}^X_{1,1}$ mode.

Figures 3.8 and 3.9 represent, again for the same dielectric waveguide, $E_y(x)$ and $H_x(x)$ along the X axis. Now the correct boundary behavior is shared by the $\mathrm{TE}^X_{1,1}$ and the $\mathrm{mTE}^X_{1,1}$ mode while the $\mathrm{TM}^Y_{1,1}$ and $\mathrm{mTM}^Y_{1,1}$ modes reveal the same type of inconsistencies as the $\mathrm{TE}^X_{1,1}$ mode along the Y axis.

From these figures we may retain the following:

- The $\mathrm{TE}^X_{1,1}$, $\mathrm{TM}^Y_{1,1}$, $\mathrm{mTE}^X_{1,1}$, and $\mathrm{mTM}^Y_{1,1}$ modes all lead to comparable field intensities inside the dielectric;
- The $\mathrm{mTE}^X_{1,1}$ mode (or the $\mathrm{mTM}^Y_{1,1}$ mode, which is very similar) appears to be the best approximate solution as far as the main transversal field components are concerned, particularly outside the dielectric where the other solutions lead to much lower field intensities, either along the X axis or along the Y axis.

The previous figures may give the impression that a significant amount of modal power is outside the dielectric, at least for the smaller waveguide cross sections. As we will show next, this is not correct and in fact most of the modal power for the $\mathrm{TE}^X_{1,1}$ (and also $\mathrm{TM}^Y_{1,1}$, $\mathrm{mTE}^X_{1,1}$, and $\mathrm{mTM}^Y_{1,1}$ modes) is concentrated inside the dielectric.

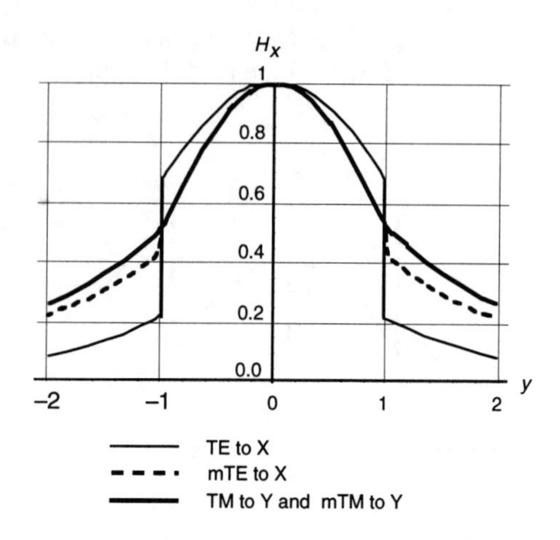

Figure 3.7: $H_x(y)$ along the Y axis for a dielectric waveguide with $\bar{\varepsilon_1} = 2.53$, $\bar{a} = 4.49$, and $\bar{b} = 1.99$.

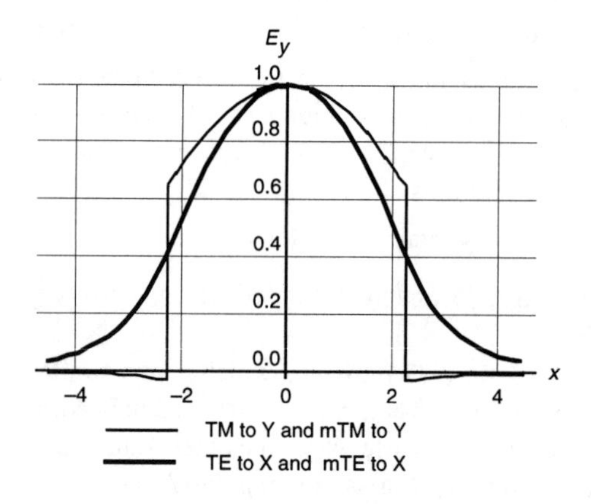

Figure 3.8: $E_y(x)$ along the X axis for a dielectric waveguide with $\bar{\varepsilon_1} = 2.53$, $\bar{a} = 4.49$, and $\bar{b} = 1.99$.

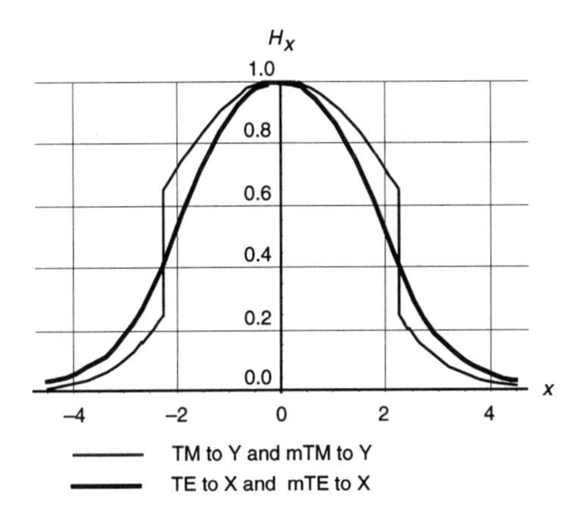

Figure 3.9: $H_x(x)$ along the X axis for a dielectric waveguide with $\bar{\varepsilon_1} = 2.53$, $\bar{a} = 4.49$, and $\bar{b} = 1.99$.

Let P_{in} and P_{out} be the modal power inside and outside the dielectric, respectively. The total modal power P_t will be

$$P_t = P_{\text{in}} + P_{\text{out}} \tag{3.119}$$

where

$$P_{\text{in}} = -\int_{-a/2}^{+a/2} dx \int_{-b/2}^{+b/2} E_{x_1} H_{y_1}^* dy \tag{3.120}$$

$$\begin{aligned} P_{\text{out}} = & -\int_{-a/2}^{-a/2} dx \int_{+b/2}^{+\infty} E_{x_2} H_{y_2}^* dy \\ & -\int_{-a/2}^{-a/2} dx \int_{-\infty}^{-b/2} E_{x_3} H_{y_3}^* dy \\ & -\int_{-\infty}^{-a/2} dx \int_{-b/2}^{+b/2} E_{x_4} H_{y_4}^* dy \\ & -\int_{+a/2}^{+\infty} dx \int_{-b/2}^{+b/2} E_{x_5} H_{y_5}^* dy \end{aligned} \tag{3.121}$$

In the above expression we have neglected the contribution of the fields in regions 6 to 9 to the modal power, in accordance to the assumptions previously made.

Taking the TE to X mode fields given in Section 3.2.2 substituting in (3.120) and (3.121), normalizing and integrating we get

$$P_{\text{in}} = \bar{\beta}(\bar{\varepsilon}_1 - \bar{k}_{x_1}^2)\left[\frac{\bar{a}}{2} + \frac{\sin(\bar{a}\bar{k}_{x_1})}{2\bar{k}_{x_1}}\right]\left[\frac{\bar{b}}{2} + \frac{\sin(\bar{b}\bar{k}_{y_1})}{2\bar{k}_{y_1}}\right] \tag{3.122}$$

$$P_{\text{out}} = 2\bar{\beta}(\bar{k}_{x_1}^2 - 1)\left[\frac{\bar{a}}{2} + \frac{\sin(\bar{a}\bar{k}_{x_1})}{2\bar{k}_{x_1}}\right]\left[\frac{\exp(-\bar{b}\bar{k}_{y_2})}{2\bar{k}_{y_0}}\right] + 2\bar{\beta}(\bar{k}_{x_0}^2 - 1)\left[\frac{\bar{a}}{2} + \frac{\sin(\bar{b}\bar{k}_{y_1})}{2\bar{k}_{y_1}}\right]\left[\frac{\exp(-\bar{a}\bar{k}_{x_0})}{2\bar{k}_{x_0}}\right] \quad (3.123)$$

Figure 3.10 shows the variation of P_{in}/P_t with $\bar{a}$ for the $\text{TE}_{1,1}^X$ mode of a dielectric waveguide with an aspect ratio of 4/9 and $\bar{\varepsilon}_1 = 2.53$, where the propagation constants were calculated according to the effective dielectric constant method. From this figure it is obvious that for $\bar{a} \geq 5$ we have $P_{\text{in}}/P_t \approx 1$.

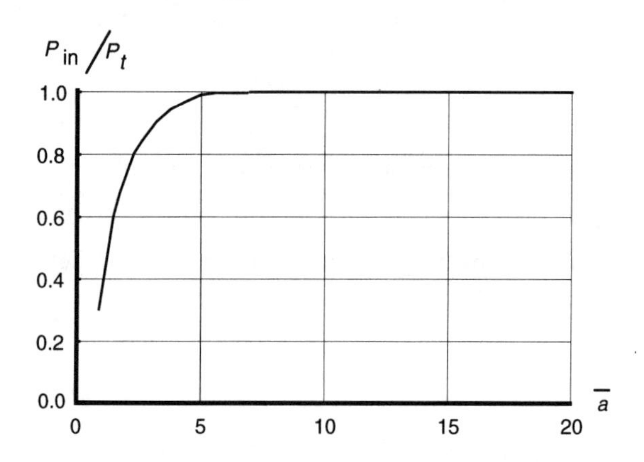

Figure 3.10: Variation of P_{in}/P_t with $\bar{a}$ for the $\text{TE}_{1,1}^X$ mode of a dielectric waveguide with an aspect ratio of 4/9 and $\bar{\varepsilon}_1 = 2.53$.

3.6 $\text{TE}_{1,1}^X$ AND $\text{TM}_{1,1}^Y$ EXCITATION EFFICIENCY

Having described the modal fields in the dielectric waveguide the next step is to determine if they can be excited, with reasonable efficiency, using as simple a launcher as possible. For this we will use a modal matching technique. Consider again the semi-infinite dielectric waveguide with ε_1 and μ_0 immersed in an otherwise free half-space with ε_0 and μ_0 and the coordinate system shown in Figure 3.1. Assume now that an electromagnetic field structure $\boldsymbol{E}_i$,$\boldsymbol{H}_i$ impinges on the left-hand side ($z < 0$) of plane XOY and that this is the only field structure that can exist on that half-space characterized by ε_0 and μ_0.

Expressing the fields on the right-hand side of plane XOY as a sum of the dielectric waveguide modes $\boldsymbol{E}_{m,n}$,$\boldsymbol{H}_{m,n}$ and matching the tangential field components at the boundary we have

$$\boldsymbol{E}_i(1 + \Gamma_r^{m,n}) = \sum_{m,n} A_{m,n}\boldsymbol{E}_{m,n} \quad (3.124)$$

$$\boldsymbol{H}_i(1 - \Gamma_r^{m,n}) = \sum_{m,n} B_{m,n}\boldsymbol{H}_{m,n} \tag{3.125}$$

where $\Gamma_r^{m,n}$ is the voltage reflection coefficient and $A_{m,n}$ and $B_{m,n}$ are the mode amplitudes.

Strictly, we should have written Γ_r instead of $\Gamma_r^{m,n}$ since it should be a constant, independent of the modes in the dielectric waveguide. However, because we assumed only one mode in the launcher, Γ_r cannot be a constant. A wide variation in the calculated values of $\Gamma_r^{m,n}$ must be taken as a warning that the above-mentioned assumption is not reasonable.

From the orthogonality of modes and (3.124) an (3.125) it follows that

$$a_{m,n} = \frac{A_{m,n}}{1 + \Gamma_r^{m,n}} = \frac{\int_S \boldsymbol{E}_i \times \boldsymbol{H}^*_{m,n} \cdot \boldsymbol{z}\, dS}{\int_S \boldsymbol{E}_{m,n} \times \boldsymbol{H}^*_{m,n} \cdot \boldsymbol{z}\, dS} \tag{3.126}$$

$$b_{m,n} = \frac{A_{m,n}}{1 - \Gamma_r^{m,n}} = \frac{\int_S \boldsymbol{E}_{m,n} \times \boldsymbol{H}^*_i \cdot \boldsymbol{z}\, dS}{\int_S \boldsymbol{E}_{m,n} \times \boldsymbol{H}^*_{m,n} \cdot \boldsymbol{z}\, dS} \tag{3.127}$$

$$\tag{3.128}$$

where S, the surface of integration, is the XOY plane, and $\boldsymbol{z}$ is the unit vector in the direction of the Z axis.

From (3.126) and (3.127) we get

$$A_{m,n} = \frac{2a_{m,n}b_{m,n}}{a_{m,n} + b_{m,n}} \tag{3.129}$$

$$\Gamma_r^{m,n} = \frac{b_{m,n} - a_{m,n}}{a_{m,n} + b_{m,n}} \tag{3.130}$$

The excitation efficiency $\eta_{m,n}$ of any given mode is now defined as the ratio between the power in that mode and the incident power:

$$\eta_{m,n} = \frac{A_{m,n}A^*_{m,n} \int_S \boldsymbol{E}_{m,n} \times \boldsymbol{H}^*_{m,n} \cdot \boldsymbol{z}\, dS}{\int_S \boldsymbol{E}_i \times \boldsymbol{H}^*_i \cdot \boldsymbol{z}\, dS} \tag{3.131}$$

Using as the launcher a rectangular waveguide operated in the $TE_{1,0}$ mode, with width a_m and the same aspect ratio (b/a) as the dielectric waveguide, we computed the excitation efficiency for the $TE^X_{1,1}$ and the $TM^Y_{1,1}$ modes of a dielectric waveguide ($\bar{\varepsilon_1} = 2.53$, $\bar{a} = 4.49$, $\bar{b} = 1.99$). The metallic waveguide was taken either empty ($\varepsilon_2 = \varepsilon_0$) or

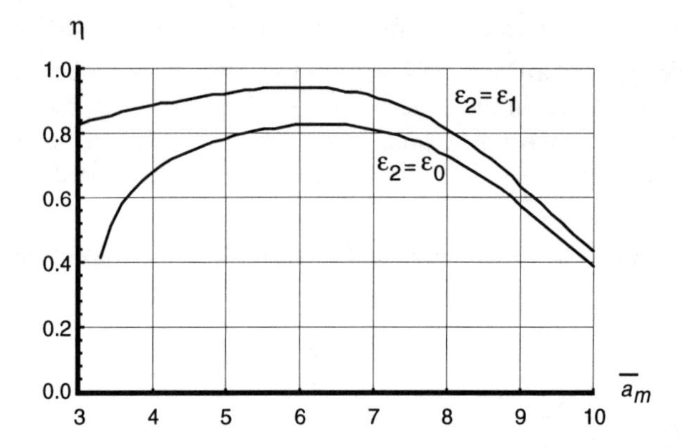

Figure 3.11: Excitation efficiency of the $TE^X_{1,1}$ mode when using a rectangular metallic waveguide either filled with the same dielectric material as the dielectric waveguide $\varepsilon_2 = \varepsilon_1$ or with no filling $\varepsilon_2 = \varepsilon_0$.

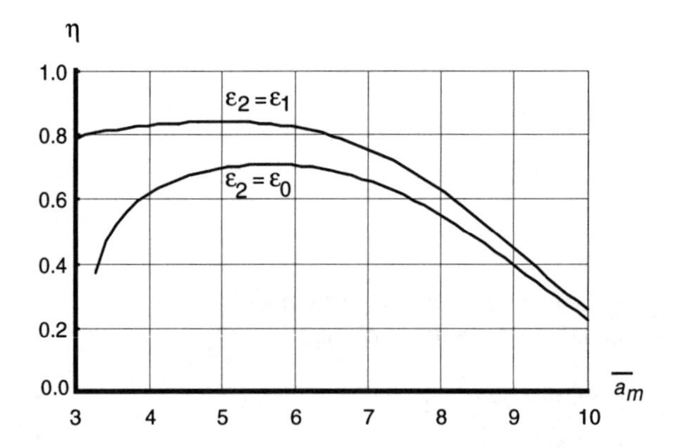

Figure 3.12: Excitation efficiency of the $TM^Y_{1,1}$ mode when using a rectangular metallic waveguide either filled with the same dielectric material as the dielectric waveguide $\varepsilon_2 = \varepsilon_1$ or with no filling $\varepsilon_2 = \varepsilon_0$.

filled with the same dielectric material as the dielectric waveguide ($\varepsilon_2 = \varepsilon_1$). The results are shown in Figure 3.11 for the $TE^X_{1,1}$ mode and in Figure 3.12 for $TM^Y_{1,1}$ mode.

In this case the excitation efficiency is quite high, particularly for the $TE^X_{1,1}$ mode. The advantage of filling the metallic waveguide with the same dielectric material as the dielectric waveguide is also obvious. Similar conclusions are obtained for the $mTE^X_{1,1}$ and $mTM^Y_{1,1}$ modes.

3.7 RADIATION PATTERN OF $TE^X_{1,1}$ MODE

The far-field radiation field $\boldsymbol{E}_P$ of a dielectric waveguide at point P may be calculated by the vector Kirchhoff-Huygens formula (2.131) that we repeat here for convenience:

$$\boldsymbol{E}_P = \frac{-jk\exp(-jkR)}{4\pi R}\boldsymbol{R}_1 \times \int_A [\boldsymbol{n} \times \boldsymbol{E} - Z\boldsymbol{R}_1 \times (\boldsymbol{n} \times \boldsymbol{H})]\exp(jk\boldsymbol{\rho} \cdot \boldsymbol{R}_1)\, dS \quad (3.132)$$

where, as before, $\boldsymbol{E}$ and $\boldsymbol{H}$ are the electric and magnetic fields on the aperture A, $\boldsymbol{R}_1$ is the unit vector from the origin to the observation point, $\boldsymbol{\rho}$ is the vector from the origin to the integration point on the aperture, and R is the distance from the origin to the observation point.

For the dielectric waveguide, surface S is formed by the "open" end or mouth of the dielectric waveguide (DCHG in Figure 3.13), the top and the bottom walls (ABCD and FEHG), and the side walls (ADGF and BCHE). Usually, when the waveguide length is considerably larger than any of its transversal dimensions, the contributions of the top, bottom, and side walls are by no means negligible when compared with the mouth.

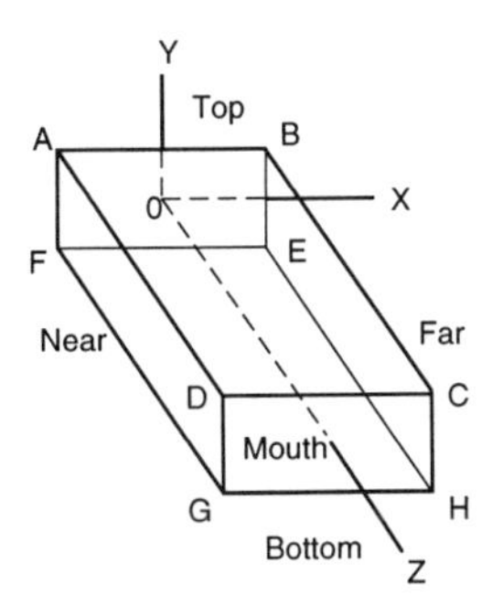

Figure 3.13: Integration surfaces for the dielectric waveguide.

The electric and magnetic fields on surface S will be obtained from the expressions derived in Section 3.2.2 assuming that the dielectric waveguide supports only the $TE^X_{1,1}$

mode. Since the transversal modal wave impedance Z_1 differs considerably from the free space impedance, at least for usual dielectric permittivities, we cannot neglect reflections at the waveguide mouth. Therefore fields will be assumed to be the result of a forward wave (moving from the excitation to the mouth) and a reflected wave traveling in the opposite direction. From (3.24), (3.27), (3.30), (3.33), (3.36), and (3.39) taking unity amplitude and normalizing all quantities we get inside the dielectric waveguide:

$$E_{x_1} = 0 \tag{3.133}$$
$$E_{y_1} = j\bar{\beta}\cos(\bar{k}_{x_1}x)\cos(\bar{k}_{y_1}y)[\exp(-j\bar{\beta}z) + \Gamma\exp(+j\bar{\beta}z)] \tag{3.134}$$
$$E_{z_1} = -\bar{k}_{y_1}\cos(\bar{k}_{x_1}x)\sin(\bar{k}_{y_1}y)[\exp(-j\bar{\beta}z) + \Gamma\exp(+j\bar{\beta}z)] \tag{3.135}$$
$$H_{x_1} = -j\frac{\bar{\varepsilon}_1 - \bar{k}_{x_1}^2}{Z_0}\cos(\bar{k}_{x_1}x)\cos(\bar{k}_{y_1}y)[\exp(-j\bar{\beta}z) - \Gamma\exp(+j\bar{\beta}z)] \tag{3.136}$$
$$H_{y_1} = -j\frac{\bar{k}_{x_1}\bar{k}_{y_1}}{Z_0}\sin(\bar{k}_{x_1}x)\sin(\bar{k}_{1y}y)[\exp(-j\bar{\beta}z) - \Gamma\exp(+j\bar{\beta}z)] \tag{3.137}$$
$$H_{z_1} = \frac{\beta\bar{k}_{x_1}}{Z_0}\sin(\bar{k}_{x_1}x)\cos(\bar{k}_{y_1}y)[\exp(-j\bar{\beta}z) + \Gamma\exp(+j\bar{\beta}z)] \tag{3.138}$$

The complex voltage reflection coefficient Γ is given by

$$\Gamma = \frac{Z_0 - Z_1}{Z_0 + Z_1}\exp(-j2\bar{\beta}\bar{l}) \tag{3.139}$$

where Z_1 is the transversal wave impedance of the field distribution in the dielectric waveguide:

$$Z_1 = \frac{E_{y_1}}{-H_{x_1}} = Z_0\frac{\bar{\beta}}{\bar{\varepsilon}_1 - \bar{k}_x^{2}} \tag{3.140}$$

and Z_0 is the free-space wave impedance.

For the dielectric waveguide mouth the normal $\boldsymbol{n}$ is in the direction of the Z axis:

$$\boldsymbol{n}_{DCHG} = \boldsymbol{z} \tag{3.141}$$

In the far-field region we have

$$\psi(x, y, z) = \exp(\boldsymbol{\rho}\cdot\boldsymbol{R}_1) = \exp[(x\cos\phi + y\sin\phi)\sin\theta + z\cos\theta] \tag{3.142}$$

where θ and ϕ define the direction along which the radiation field is calculated.

Substituting in (2.131) the values for $\boldsymbol{n}$ and ψ and applying the electric and magnetic fields given in (3.133) through (3.138) we find that, omitting a constant factor $\frac{jk\exp(-jkR)}{4\pi R}$ the contribution of the mouth is given by

$$\begin{aligned} \boldsymbol{Er}_{\text{DCHG}}(\theta, \phi) &= \int_{-\bar{a}/2}^{+\bar{a}/2} dx \\ & \int_{-\bar{b}/2}^{+\bar{b}/2} [-Z_0 ha_1 \boldsymbol{\theta} - Z_0 ha_2 \boldsymbol{\phi} + ea_1 \boldsymbol{\theta} + ea_2 \boldsymbol{\phi}]\, \psi(x, y, z = \bar{l})\, dy \end{aligned} \tag{3.143}$$

where

$$ha_1 = [H_{x_1}(z = \bar{l}) \sin\phi - H_{y_1}(z = \bar{l}) \cos\phi] \cos\theta \tag{3.144}$$

$$ha_2 = H_{x_1}(z = \bar{l}) \cos\phi + H_{y-1}(z = \bar{l}) \sin\phi \tag{3.145}$$

$$ea_1 = E_{x_1}(z = \bar{l}) \cos\phi + E_{y_1}(z = \bar{l}) \sin\phi \tag{3.146}$$

$$ea_2 = [E_{y_1}(z = \bar{l}) \cos\phi - E_{x_1}(z = \bar{l}) \sin\phi] \cos\theta \tag{3.147}$$

For the top boundary (ABCD) the unit normal is

$$\boldsymbol{n}_{\text{ABCD}} = \boldsymbol{y} \tag{3.148}$$

where as for the bottom boundary (FEHG) it is

$$\boldsymbol{n}_{\text{FEHG}} = -\boldsymbol{y} \tag{3.149}$$

Proceeding as for the mouth we get now for the top and bottom boundaries

$$\begin{aligned} \boldsymbol{Er}_{\text{ABCD}}(\theta, \phi) &= \int_{-\bar{a}/2}^{+\bar{a}/2} dx \\ & \int_0^{\bar{l}} (-Z_0 ht_1 \boldsymbol{\theta} + Z_0 ht_2 \boldsymbol{\phi} - et_1 \boldsymbol{\theta} - et_2 \boldsymbol{\phi})\, \psi(x, y = \bar{b}/2, z)\, dz \end{aligned} \tag{3.150}$$

$$\begin{aligned} \boldsymbol{Er}_{\text{FEHG}}(\theta, \phi) &= \int_{-\bar{a}/2}^{+\bar{a}/2} dx \\ & \int_0^{\bar{l}} (+Z_0 hb_1 \boldsymbol{\theta} - Z_0 hb_2 \boldsymbol{\phi} + eb_1 \boldsymbol{\theta} - eb_2 \boldsymbol{\phi})\, \psi(x, y = -\bar{b}/2, z)\, dz \end{aligned} \tag{3.151}$$

where

$$ht_1 = H_{x-1}(y = +\bar{b}/2) \sin\theta + H_{z_1}(y = +\bar{b}/2) \cos\theta \cos\phi \tag{3.152}$$

$$ht_2 = H_{z_1}(y = +\bar{b}/2) \sin\phi \tag{3.153}$$

$$et_1 = E_{z_1}(y = +\bar{b}/2)\sin\phi \tag{3.154}$$
$$et_2 = E_{x_1}(y = +\bar{b}/2)\sin\theta + E_{z_1}(y = +\bar{b}/2)\cos\theta\cos\phi \tag{3.155}$$
$$ht_1 = H_{x_1}(y = -\bar{b}/2)\sin\theta + H_{z_1}(y = -\bar{b}/2)\cos\theta\cos\phi \tag{3.156}$$
$$ht_2 = H_{z_1}(y = -\bar{b}/2)\sin\phi \tag{3.157}$$
$$et_1 = E_{z_1}(y = -\bar{b}/2)\sin\phi \tag{3.158}$$
$$et_2 = E_{x_1}(y = -\bar{b}/2)\sin\theta + E_{z_1}(y = -\bar{b}/2)\cos\theta\cos\phi \tag{3.159}$$

Similarly, for the near (ADGF) and far (EBCH) walls we have

$$n_{\mathrm{ADGF}} = -\boldsymbol{x} \tag{3.160}$$
$$n_{\mathrm{EBCH}} = \boldsymbol{x} \tag{3.161}$$

and the corresponding contributions for the radiated field

$$\boldsymbol{Er}_{\mathrm{ADGF}}(\theta,\phi) = \int_{-\bar{b}/2}^{+\bar{b}/2} dy \int_0^{\bar{l}} [-Z_0 hn_1\boldsymbol{\theta} - Z_0 hn_2\boldsymbol{\phi} + en_1\boldsymbol{\theta} - en_2\boldsymbol{\phi}]\,\psi(x = -\bar{a}/2, y, z)\,dz \tag{3.162}$$

$$\boldsymbol{Er}_{\mathrm{BCHE}}(\theta,\phi) = \frac{jk_0}{4\pi}\int_{-\bar{b}/2}^{+\bar{b}/2} dy \int_0^{\bar{l}} [+Z_0 hf_1\boldsymbol{\theta} + Z_0 hf_2\boldsymbol{\phi} - ef_1\boldsymbol{\theta} + ef_2\boldsymbol{\phi}]\,\psi(x = +\bar{a}/2, y, z)\,dz \tag{3.163}$$

where

$$hn_1 = H_{y_1}(x = -\bar{a}/2)\sin\theta + H_{z_1}(x = -\bar{a}/2)\cos\theta\sin\phi \tag{3.164}$$
$$hn_2 = H_{z_1}(x = -\bar{a}/2)\cos\phi \tag{3.165}$$
$$en_1 = E_{z_1}(x = -\bar{a}/2)\cos\phi \tag{3.166}$$
$$en_2 = E_{y_1}(x = -\bar{a}/2)\sin\theta + E_{z_1}(x = -\bar{a}/2)\cos\theta\sin\phi \tag{3.167}$$
$$hf_1 = H_{y_1}(x = +\bar{a}/2)\sin\theta + H_{z_1}(x = +\bar{a}/2)\cos\theta\sin\phi \tag{3.168}$$
$$hf_2 = H_{z_1}(x = +\bar{a}/2)\cos\phi \tag{3.169}$$
$$ef_1 = E_{z_1}(x = +\bar{a}/2)\cos\phi \tag{3.170}$$
$$ef_2 = E_{y_1}(x = +\bar{a}/2)\sin\theta + E_{z_1}(x = +\bar{a}/2)\cos\theta\sin\phi \tag{3.171}$$

Substituting in the above formulas the E- and H-field components by their values given in (3.133) through (3.138) and taking successively $\phi = \pi/2$ and $\phi = 0$ we obtain the E- and H-plane far-field radiation patterns. In this case the integrals have been performed in closed form using *Mathematica* [19]. The resulting formulas are, however, too long to be included in the text.

The E- and the H-plane far-field computed radiation patterns due to the mouth, the top and bottom walls, the near and far side walls and, finally, the overall radiation pattern, of the $TE^X_{1,1}$ mode in dielectric waveguide with $\bar{\varepsilon}_1 = 2.53$, $\bar{a} = 4.49$, $\bar{b} = 1.99$, and $\bar{L} = 16.61$, excited by an X-band metallic waveguide operated in the $TE_{1,0}$ mode are depicted in Figure 3.14. As shown the top and bottom walls play a very important role in shaping the radiation pattern, both increasing its directivity and bringing in sidelobes which are mostly absent in the radiation pattern of the mouth alone.

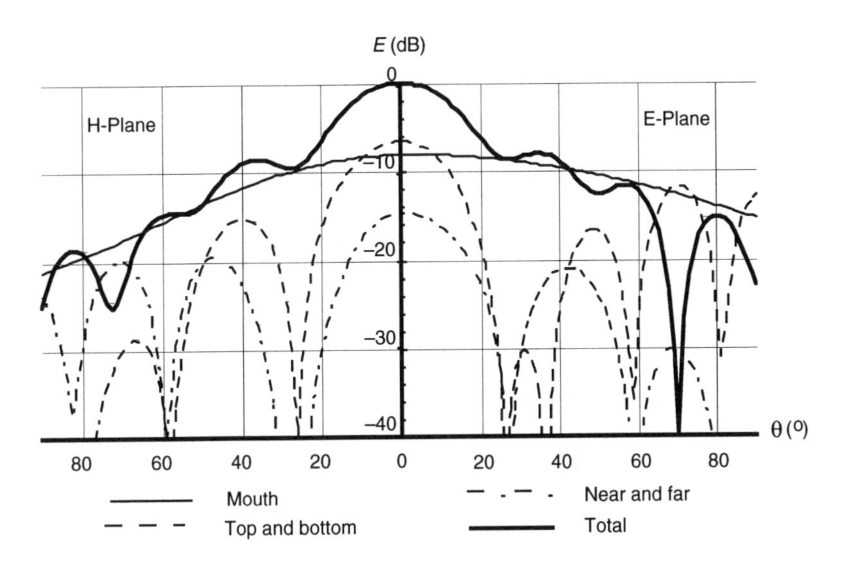

Figure 3.14: Calculated E- and H-plane radiation patterns of the $TE^X_{1,1}$ mode of a dielectric waveguide with $\bar{\varepsilon}_1 = 2.53$, $\bar{a} = 4.49$, $\bar{b} = 1.99$, and $\bar{L} = 16.61$, showing the contribution of the mouth, the top and bottom, and the near and far walls.

The overall computed and measured radiation patterns are shown in Figure 3.15. The agreement between the computed and measured patterns is is good on the main lobe but deteriorates somewhat on the sidelobes. This is to be expected since it is at the sidelobe level that the influence of the radiation pattern of the metallic waveguide is more likely to be felt.

Since the radiation pattern is the result of contributions from five surfaces we should not expect a well-defined phase center. This is indeed the case. Such behavior discourages the use of the dielectric waveguide as a feeder in conjunction with a reflector.

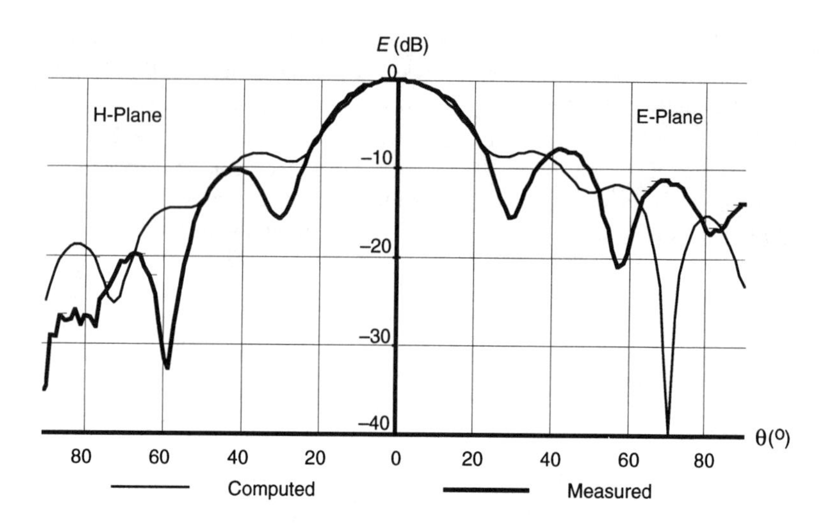

Figure 3.15: Calculated and measured E- and H-plane radiation patterns for the $TE^X_{1,1}$ mode of a dielectric waveguide with $\bar{\varepsilon_1} = 2.53$, $\bar{a} = 4.49$, $\bar{b} = 1.99$, and $\bar{L} = 16.61$.

3.8 EFFECT OF LAUNCHER

As launching efficiency η is always less than unity, one should expect the launcher to influence the measured radiation pattern. This effect should be more pronounced for lower efficiencies and should be more apparent at the sidelobe level where the radiated fields due to the dielectric waveguide are not likely to dominate those due to the launcher.

The rigorous analysis of the effect of the launcher is a very complex problem which will not be attempted here. There is, however, a simple alternative put forward by James[20], which assumes the launcher to radiate unaffected by the presence of the dielectric, except for a reduction in power being converted into dielectric modes. Thus, in the far field we should have

$$E_t(\theta, \phi) = E_d(\theta, \phi) + E_l(\theta, \phi) \tag{3.172}$$

where E_t is the total field resulting from the dielectric waveguide E_d and the launcher E_l.

Contributions E_d and E_l may be obtained from the respective radiation patterns normalized to boresight values:

$$\bar{E}_d(\theta, \phi) = \frac{E_d(\theta, \phi)}{E_d(\theta = 0, \phi = 0)} \tag{3.173}$$

$$\bar{E}_l(\theta, \phi) = \frac{E_l(\theta, \phi)}{E_l(\theta = 0, \phi = 0)} \tag{3.174}$$

From the boresight gains of the dielectric antenna g_d and of the launcher g_l and the excitation efficiency η it follows that

$$\frac{E_l(\theta = 0, \phi = 0)}{E_d(\theta = 0, \phi = 0)} = \sqrt{\frac{1-\eta}{\eta}\frac{g_l}{g_d}} \tag{3.175}$$

Dividing (3.172) by $E_d(\theta = 0, \phi = 0)$ substituting the value of $\frac{E_l(\theta=0,\phi=0)}{E_d(\theta=0,\phi=0)}$ given in (3.175) and manipulating we get

$$E_t(\theta, \phi) = E_d(\theta = 0, \phi = 0)\left[\bar{E}_d(\theta, \phi) + \sqrt{\frac{1-\eta}{\eta}\frac{g_l}{g_d}}\bar{E}_l(\theta, \phi)\right] \tag{3.176}$$

For display, the resulting radiation pattern should normally be normalized to boresight, that is, divided by $E_t(\theta = 0, \phi = 0)$.

Before adding the launcher contribution we must ensure that the field values used to derive the launcher and the dielectric waveguide patterns are compatible, that is, they have the same phase references, and that both radiation patterns are referred to the same point, here taken to be the origin of the coordinates.

Applying this method to the previous example, of a dielectric waveguide with $\bar{\varepsilon}_1 = 2.53$, $\bar{a} = 4.49$, $\bar{b} = 1.99$, and $\bar{L} = 16.61$, excited by an X-band metallic waveguide at 9.375 GHz and taking $\eta = 0.9$ and $g_d = 17.98$ and $g_l = 4.64$ we obtained the overall radiation patterns shown in Figure 3.16 together with the measured radiation patterns.

The value of g_d was obtained from (2.220) and (2.221) using an antenna efficiency of 100%, that is, neglecting dielectric losses.

In order to evaluate the accuracy of the method and since radiation patterns of small aperture such as the rectangular metallic waveguide are rather difficult to predict accurately, using simple expressions, we used the measured radiation pattern of the launcher (Figure 3.17) and calculated g_l by pattern integration, which again is equivalent of taking the directivity for the gain, a reasonable approximation for these types of antennas.

As can be seen comparing Figures 3.15 and 3.16 including the effect of the launcher has a limited impact on the quality of the predictions.

Much better results may be obtained if we modify the launcher phase pattern to include the presence of the dielectric waveguide. This effect is assumed to be equivalent to the propagation delay introduced by a dielectric path $d\cos\theta$ where d equals the dielectric waveguide length L and θ is the angle from boresight (Z axis). If the normalized propagation constant in this dielectric path is taken as $\bar{\beta}_{da}$, then, in the far-field region, this delay is equivalent to a displacement Δz of the center of rotation of the launcher, in the negative direction of the Z axis:

$$\Delta z = L(\bar{\beta}_{da} - 1) \tag{3.177}$$

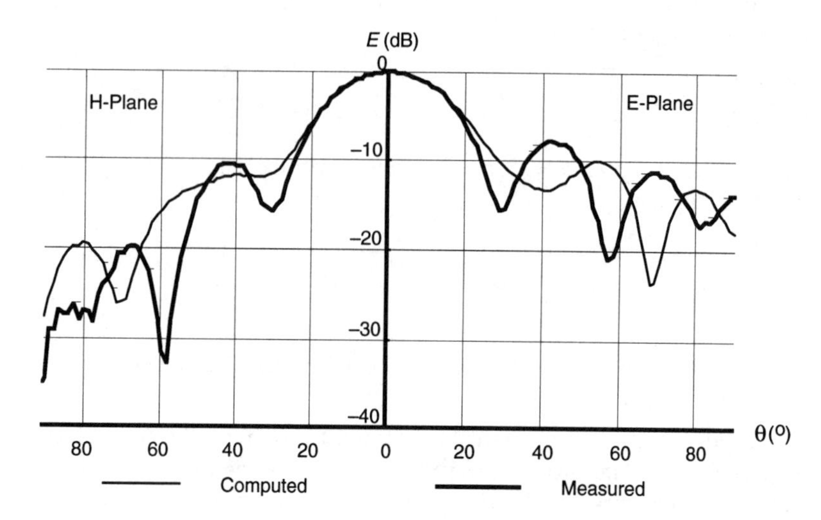

Figure 3.16: Computed, according to James's method [20], and measured E-plane radiation patterns for the $TE^{X}_{1,1}$ mode of a dielectric waveguide with $\bar{\varepsilon}_1 = 2.53$, $\bar{a} = 4.49$, $\bar{b} = 1.99$, and $\bar{L} = 16.61$.

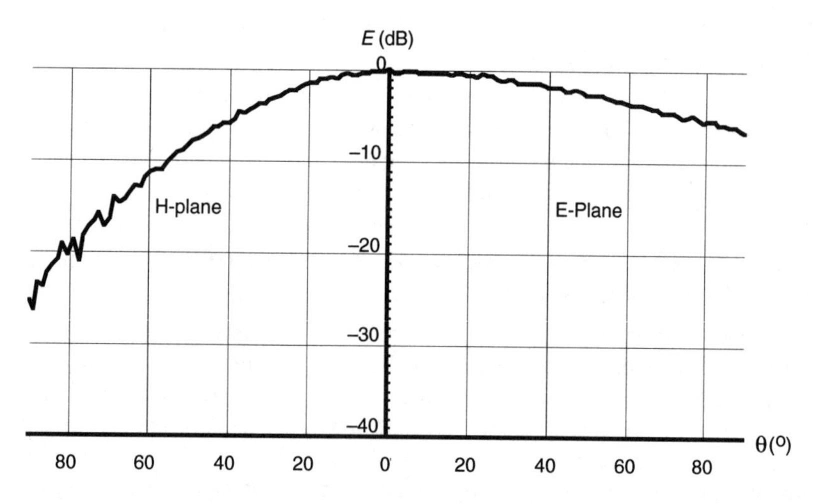

Figure 3.17: Measured E- and H-plane radiation patterns of a standard X-band metallic waveguide ($\bar{a} = 4.49$ and $\bar{b} = 1.99$) filled with dielectric $\bar{\varepsilon_1} = 2.53$.

The difficulty in applying (3.177) is that we do not known the value of $\bar{\beta}_{da}$ even if we can bound it $1 \leq \bar{\beta}_{da} \leq \sqrt{\bar{\varepsilon}_1}$. Experience shows that fairly good results are obtained using a value of Δz given by

$$\Delta z = k_p L \sqrt{\bar{\beta} - 1} \tag{3.178}$$

where $\bar{\beta}$ is the normalized longitudinal propagation constant in the dielectric waveguide and the value of k_p, depends on the length and shape of the dielectric structure but is often in the range of 0.65 to 0.75.

Assuming the launcher phase pattern is measured about its phase center, as is usually the case, one has first to consider the effect of moving the center of rotation to the origin of the coordinate system, centered on the metallic waveguide aperture (Figure 3.1), and in addition to introduce the equivalent displacement of the center of rotation of the launcher.

Figure 3.18 shows the measured and computed radiation patterns of a dielectric waveguide with $\bar{\varepsilon_1} = 2.53$, $\bar{a} = 4.49$, $\bar{b} = 1.99$, and $\bar{L} = 16.61$, excited by an X-band metallic waveguide at 9.375 GHz, using the method just described with $g_l = 4.64$, $g_d = 16.71$, $k_p = 0.75$, and $\eta = 0.9$. The agreement between the two patterns is now significantly improved over the previous results and extends well beyond the main lobe.

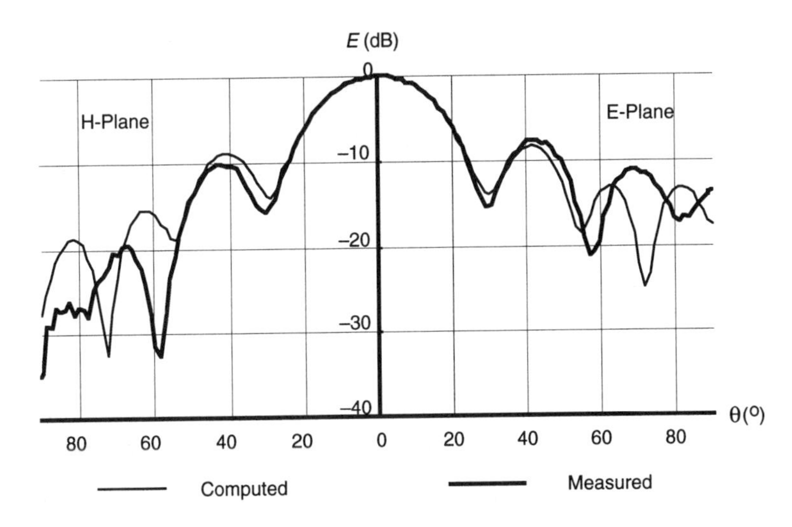

Figure 3.18: Calculated, according to the modified James's method, and measured E- and H-plane radiation patterns for the $TE^X_{1,1}$ mode of a dielectric waveguide with $\bar{\varepsilon}_1 = 2.53$, $\bar{a} = 4.49$, $\bar{b} = 1.99$, and $\bar{L} = 16.61$.

3.9 SHIELDED DIELECTRIC WAVEGUIDE

As presented above, the dielectric waveguide is of little use as a practical antenna mainly because of the high sidelobe level. In addition, as the radiation pattern results from five radiating surfaces plus the launcher there is no well-defined phase center.

Sidelobes are mainly due to the lateral walls and launcher (see Figures 3.14 and 3.15). By itself the mouth pattern is rather wide, particularly in the E-plane, virtually sidelobe free with a phase center on the mouth.

It would therefore be most interesting to find out a way to eliminate all unwanted components in the radiation pattern. This may be easily achieved by inserting the dielectric waveguide antenna coaxially into a suitably sized metallic shield (Figure 3.19) that results in the shielded dielectric waveguide antenna. If the dimensions of the metallic shield cross section are such that the dielectric field modes are negligible (about 20 dB below their maximum value) on the shield, the metallic walls have little or no effect on the dielectric waveguide field structure. Nevertheless they block radiation from the side walls and suppress launcher effects, and the radiation pattern becomes that of dielectric waveguide free end and surrounding space. The latter is particularly important for small dimension waveguide cross sections.

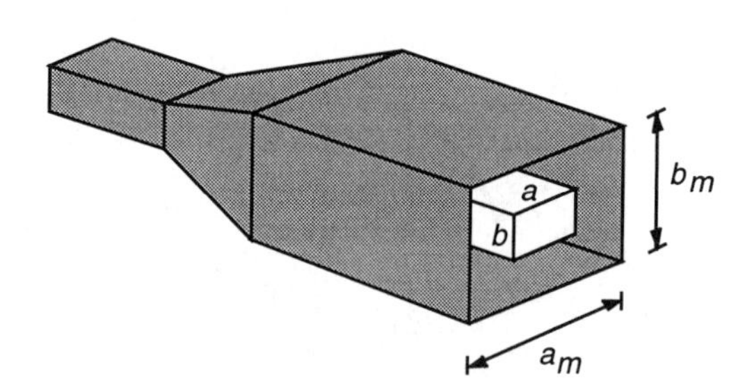

Figure 3.19: Shielded dielectric waveguide antenna.

The use of an absorbent shield was suggested by Clarricoats and Salema [21] who applied it to reduce the sidelobe level of a dielectric cone antenna excited by a metallic pyramidal horn. A dielectric loaded metallic horn was first put forward by Clarricoats et al. [22] and later by Lier [23, 24] for a circular cross section. Later still a rectangular cross section was used by Lier [25]. While in the former the absorbent shield, like the metal shield in the shielded dielectric waveguide, has a negligible interaction with the dielectric field structure, in the latter the opposite is true.

The most convenient way to compute the radiation pattern for the shielded dielectric waveguide antenna is to use the Kirchhoff-Huygens formulas using a plane aperture limited by the shield. Unlike the previous case this requires an accurate description of the electromagnetic fields outside the dielectric which is all the more important for small (i.e., $< 0.7\lambda$) dielectric cross section dimensions where the fields extend well outside the dielectric. In addition, neglecting the fields in regions 6 through 9 (Figure 3.2), as suggested in Section 3.2.1, leads to a considerable underestimation of the directivity. Thus, in this case we take the following approach:

- Use the mTE to X to improve field description outside the dielectric;
- Calculate the transversal propagation constants using the "effective dielectric constant" method;
- Describe the fields in regions 6 through 9 heuristically by

$$E_{y_{6,7,8,9}} = +j\beta\zeta\xi \exp(-k_{x_0}|x|)\exp(-k_{y_0}|y|) \tag{3.179}$$

Figure 3.20 illustrates the resulting computed and measured radiation patterns for a dielectric waveguide with $\bar{\varepsilon}_1 = 2.53$, $\bar{a} = 4.49$, $\bar{b} = 1.99$, and $\bar{L} = 16.61$, excited by an X-band metallic waveguide at 9.375 GHz and inserted into a metallic rectangular shield with aperture cross section dimensions $\bar{a}_m = 11.58$ and $\bar{b}_m = 12.57$. The remarkable agreement between computed and measured patterns justifies the approach adopted.

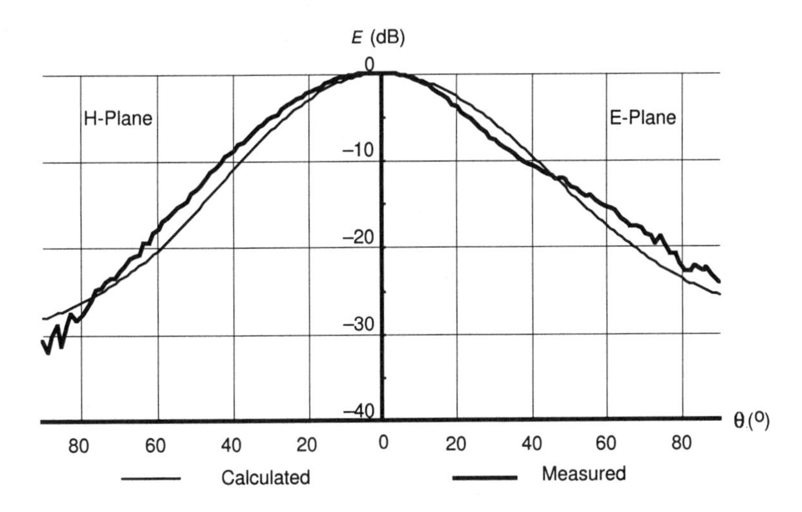

Figure 3.20: Calculated and measured E- and H-plane radiation patterns for the $\text{mTE}^X_{1,1}$ mode of a shielded dielectric waveguide with $\bar{\varepsilon}_1 = 2.53$, $\bar{a} = 4.49$, $\bar{b} = 1.99$, $\bar{a}_m = 11.58$, and $\bar{b}_m = 12.57$.

The shielded dielectric waveguide radiation pattern shows very convenient features: low sidelobe level and appreciable circular symmetry. Measurements show that it has a low cross polarization level and a well-defined phase center over a fairly wide bandwidth. Waveguide to antenna impedance matching is acceptable for some applications (measured reflection level between -15 and -20 dB) and may be improved by simple matching devices.

Dielectric waveguide antenna geometry, by using equal (or at least similar) launcher and dielectric cross sections, places unduly restrictions on the design of an antenna to a prescribed radiation pattern. Horns, dealt with in Chapter 4, provide the required extra degrees of freedom.

3.10 CONCLUSIONS

The fields of an infinite dielectric waveguide were described in terms of simple trigonometric functions inside the dielectric and decaying exponential functions outside. Such a simple description, due to Marcatili [6], is only an approximate solution of Maxwell's equation for such a structure since not all boundary conditions can be simultaneously met. However, it works well particularly far from the cutoff region and has, over other available solutions, the considerable advantage of simplicity. In addition, its range of application may be considerably extended near the cutoff region and its accuracy improved, while keeping the simplicity, with the "effective dielectric constant" method. The transversal and the axial propagation constants calculated by both methods are represented in a new diagram (Figure 3.5) that highlights the range where both methods yield comparable results.

If we concentrate on the transversal, rather than the longitudinal, field components then, for small dimension dielectric cross sections, Marcatili's characteristic equations lead to a significant underestimation of the fields outside the dielectric.

Based on Marcatili's field descriptions for the TE to X and the TM to Y modes we derived an alternative set of characteristic equations, common to both modes, which leads to a more realistic behavior of the transversal field components at the dielectric boundary and which, while yielding comparable field values inside the dielectric, provides more accurate predictions of the field intensities outside the dielectric

Due to the similarity between the fields of the $TE_{1,0}$ mode of a metallic rectangular waveguide and the TE to X and the TM to Y modes of a dielectric waveguide, high values of the excitation efficiency can be achieved using the former as the launcher.

Marcatili's description of the electromagnetic fields, modified by the concept of the "effective dielectric constant," was used to explain and predict the far-field radiation pattern of a dielectric waveguide antenna. The Kirchhoff-Huygens formula was applied to an aperture that included not only the dielectric waveguide free end, here known as the mouth, but also the lateral walls: top, bottom, near, and far.

The launcher was assumed to radiate unaffected by the presence of the dielectric waveguide, except for a reduction in power being converted into dielectric modes and a shift in the center of rotation, in the negative direction of the Z axis, basically depending on dielectric length and permittivity. These assumptions enabled us to add the effect of the launcher pattern to the dielectric waveguide pattern and to predict the overall radiation pattern with excellent accuracy.

The structure of the dielectric waveguide radiation pattern is heavily influenced by the lateral walls, as well as by the launcher pattern. The radiation pattern is reasonably symmetric on the main lobe but a high sidelobe level detracts from this symmetry and is likely to be a considerable handicap in many applications

Enclosing the dielectric waveguide in a metallic shield with cross section dimensions such that on it the dielectric waveguide modal fields are negligible in such a way that only the waveguide free end radiates provides a simple and inexpensive way to eliminate side wall and launcher effects thus providing an almost sidelobe level free pattern. This approach differs from the dielectric loaded horns in that the metal is used only as a shield and does not affect the dielectric fields to a significant extent.

Expanding the dielectric waveguide cross section toward the dielectric waveguide free end provides a way to increase the radiating aperture so as to match a required radiation pattern. This will be the subject of Chapter 4.

REFERENCES

[1] Southworth, G. C., (1940), U.S. Patent 2,206,923.

[2] Mueller, G. E., and Tyrrell, W. A., (1947), Polyrod Antennas, *Bell System Technical Journal*, Vol. 26, pp. 837–851.

[3] Shiau, Y. (1976), Dielectric Rod Antennas for Millimeter-Wave Integrated Circuits, *IEEE Transactions on Microwave Theory and Techniques*, Vol. MTT-24, No. 11, pp. 869–872.

[4] Yao, C., Schwarz, S. E., and Blumenstock, B. J., (1982), Monolithic Integration of a Dielectric Millimeter-Wave Antenna and Mixer Diode: An Embryonic Millimeter-Wave IC, *IEEE Transactions on Microwave Theory and Techniques*, Vol. MTT-30, No. 8, pp. 1241–1247.

[5] Watson, R. B. and Horton, C. W., (1948), The Radiation Patterns of Dielectric Rods — Experiment and Theory, *Journal of Applied Physics*, Vol. 19, pp. 661–670.

[6] Marcatili, E.A.C., (1969), Dielectric Rectangular Wave Guide And Directional Coupler for Integrated Optics, *Bell System Technical Journal*, Vol. 48, No. 7, pp. 2071 –2102.

[7] McLevige, W. V., Itoh, T., and Mittra, R., (1975), New Waveguide Structures for Millimeter-Wave and Optical Integrated Circuits, *IEEE Transactions on Microwave Theory and Techniques*, Vol. MTT-23, No. 10, pp. 788–794.

[8] Goell, J. E., (1969), A Circular Harmonic Computer Analysis of Rectangular Dielectric Waveguide, *Bell System Technical Journal*, Vol. 48, No. 7, pp. 2133 – 2160.

[9] Eyges, L., Gianino, P. and Wintersteiner, P. (1979), Modes of Dielectric Waveguides of Arbitrary Cross Sectional Shapes, *Journal of the Optical Society of America*, Vol. 69, No.9, pp. 1226 – 1235.

[10] Morita, M., (1982), A Method of Extending The Boundary Conditions For Analyzing Guided Modes of Dielectric Wave Guides of Arbitrary Cross-Sectional Shape, *IEEE Transactions on Microwave Theory and Techniques*, MTT-30, No. 1, pp. 6–13.

[11] Sen, T. K., and Chatterjee, R., (1978), Rectangular Dielectric Rod at Microwave Frequencies — Part I. Theoretical and Experimental Determination of Launching Efficiency, *Journal of the Indian Institute of Science*, Vol. 60, No. 5, pp. 193–210.

[12] Sen, T. K., and Chatterjee, R., (1978), Rectangular Dielectric Rod at Microwave Frequencies — Part II. Radiation Characteristics, *Journal of the Indian Institute of Science*, Vol. 60, No. 5, pp. 211–225.

[13] Paulit, S. K., and Chatterjee, R., (1983), Radiation Characteristics of Tapered Rectangular Dielectric Rod Antenna at Microwave Frequencies, *Proceedings of the 1983 URSI Symposium on Electromagnetic Theory*, Santiago de Compostela, Spain, pp. 401–404.

[14] Kobayashi, S., Mittra, R., and Lampe, R., (1982), Dielectric Tapered Rod Antennas for Millimeter-Wave Applications, *IEEE Transactions on Antennas and Propagation*, Vol. AP-30, No. 1, pp. 54–58.

[15] Zucker, F. J., (1961), Surface and Leaky Wave Antenna, in Jasik, H. (ed.), *Antenna Engineering Handbook*, Jasik, H. (Ed), McGraw-Hill, New York.

[16] Sudbø, A. S., (1992), Why Are Accurate Computations of Mode Fields in Rectangular Waveguides Difficult?, *Journal of Lightwave Technology*, Vol. 10, No. 4, pp. 418–419. .

[17] Harrington, R. F., (1961), *Time Harmonic Electomagnetic Fields*, McGraw-Hill, New York.

[18] Chatterjee, R., (1985), *Dielectric and Dielectric-Loaded Antennas*, Research Studies Press, Letchworth, Hertfordshire, England.

[19] Wolfram Research Inc., (1995), *Mathematica*, Version 2.2, Wolfram Research Inc.

[20] James, J. R., (1967), *Studies of Cylindrical Dielectric Antennas*, Royal Military College of Science, Department of Electrical and Instrument Technology, Technical Note RT41.

[21] Clarricoats, P. J. B., and Salema, C. E. R. C., (1972), Influence of Launching Horn on Radiation Characteristics of a Dielectric Cone Feed, *Electronics Letters*, Vol. 8, No. 8, pp. 200-202.

[22] Clarricoats, P. J. B., Olver, A. D., and Rizk, M. S. A. S., (1983), A Dielectric Loaded Conical Feed with low cross-polar radiation, *Proceedings of the URSI Symposium on Electromagnetic Theory*,Santiago de Compostela, Spain, pp. 351–354.

[23] Lier, E., and Aas, J. A., (1985), Simple Hybrid Mode Horn Feed Loaded with a Dielectric Cone, *Electronics Letters*, Vol. 21, No. 13, pp. 563–564.

[24] Lier, E., (1986), A Dielectric Hybrid Mode Antenna Feed: A Simple Alternative to the Corrugated Horn, *IEEE Transactions on Antennas and Propagation*, Vol. AP-34, No. 1, pp. 21–29.

[25] Lier, E., and Stoffels, C., (1991), Propagation and Radiation Characteristics of Rectangular Dielectric-loaded Hybrid Mode Horn, *Proceedings of the IEE (H)*, Vol. 138, No. 5, pp. 407–411.

Chapter 4

Dielectric Sectoral and Pyramidal Horns

4.1 INTRODUCTION

As discussed in Section 3.1, the first works on rectangular cross section dielectric antennas appear to be due to Mueller [1] in 1947 and to Wilkes [2] and Watson [3] in 1948. Later workers, such as Shiau [4] and Kobayashi [5] have exploited tapering down the dielectric, that is, reducing its cross section toward the free end to improve the radiation characteristics as end-fire antennas. Paulit and Chatterjee [6] have extensively studied the radiation properties of the tapered rectangular dielectric rod antenna, with very low semiflare angles (up to -1.97 degrees).

Solid dielectric horn antennas, as opposed to tapered antennas, appear to have attracted far less attention from researchers. Ohtera [7] presented the radiation characteristics without theoretical details of a modified rhombic dielectric plate antenna which could be described as a dielectric H-plane sectoral horn modified by the addition of a simple triangular-shaped lens. Jha [8] and Singh [9] derived the far-field radiation patterns of H- and E-plane solid dielectric sectoral horns, respectively, using the vector Kirchhoff-Huygens formulas, but failed to consider aperture phase error, reflections from the mouth, dependence of the dielectric fields on the horn's variable cross section, and launcher effects. Aperture fields computed using Marcatili's dielectric waveguide approach were found to be in good agreement with measurements [10, 11]. In 1974, Brooking [12] reported that a low-permittivity ($\bar{\varepsilon} \approx 1.05$) pyramidal horn, excited by a metallic pyramidal horn, showed increased gain, lower sidelobe level, and improved circular symmetry when compared to the launcher alone. The radiation pattern was predicted using the two-aperture theory referred to in Chapter 3, where the apertures are the launcher and the pyramidal horn mouth. The fields on the dielectric are assumed to be the same as in a dielectric rod with the same cross section [20] plus a phase error, accounting for the varying propagation velocity and path length.

At microwave frequencies the metallic pyramidal horn is a well-known low-cost antenna, with excellent cross-polarization discrimination and reasonably high-aperture efficiency, but its radiation pattern may fall short on the sidelobe level (particularly in the E-plane). A variety of solutions have been suggested of which a few are worth recalling:

- Using (two) internal metallic plates, normal to the electric field, to introduce an amplitude tapering of the aperture field in the E-plane [13];
- Placing two-high impedance (about $\lambda/4$) choke flanges, at the horn aperture edges normal to the electric field, with the slots facing forward [14, 15];
- Corrugating, up to about $\lambda/4$ deep, the two horn internal walls normal to the electric field [14, 16];
- Loading the horn with dielectric slabs, normal to the electric field and adjacent to the edge walls [17];
- Combining internal metallic plates, normal to the electric field, with dielectric slabs near the edges and parallel to the electric field [18];
- Introducing a central dielectric slab normal to the electric field [19].

More recently Lier [21] suggested the dielectric loaded pyramidal horn, a dielectric pyramidal horn enclosed in a suitably sized metallic pyramidal horn as a means of obtaining very low sidelobes coupled with a low cross-polarization level and good circular symmetry.

In this chapter we will deal with the radiation properties of solid dielectric horns of low to moderate (up to 45 degrees) flare angle.

We will start by showing how the solid dielectric sectoral and pyramidal horn shapes derive from the dielectric waveguide in the manner often used for their metallic counterparts.

Similar to the dielectric waveguide there is no exact solution of Maxwell's equations for solid dielectric horns. An approximate method, based on the dielectric waveguide formulation presented in Chapter 3, is used to derive expressions for the electromagnetic fields.

The radiation patterns of the solid dielectric horns is obtained as a superposition of the radiation patterns of a finite set of elementary waveguides of gradually expanding cross section. The result is calculated as a sum of the contributions of the top, bottom, side, and step walls of each elementary waveguide plus the radiated fields due to the mouth of the last (largest) elementary waveguide. The fields inside each elementary waveguide are taken as the sum of a forward wave plus a backward wave due to reflections at the mouth. The field distribution for each wave is the unperturbed field of an infinite dielectric waveguide with the same cross section and constitutive parameters. The relative field amplitudes in successive elementary waveguides are adjusted so as to be consistent with power conservation.

The overall radiation patterns of solid dielectric horns are usually influenced to a significant extent by the launcher. Here, this effect is evaluated in a manner similar to the one described by James [22]. The launcher radiation is assumed unaffected by the presence of the dielectric horn, except for a reduction in the radiated power by $(1 - \eta)$ where η is the excitation efficiency and the introduction of a phase delay depending on the horn length and permittivity. The overall pattern is obtained by vector addition of the patterns of the dielectric horn and the launcher, with due consideration to the excitation efficiency. Measurements with a number of horns up to 45-degree flare angle confirm the approach taken and justify the decision not to use more elaborate analysis.

Inserting the solid dielectric horn in a metallic shield, in a manner akin to the one described in Section 3.9 for the dielectric waveguide, we obtain a shielded solid dielectric horn. The shield considerably reduces the contribution of the sidewalls and launcher to the radiation pattern and the end result is an almost sidelobe-free pattern. Compared to the shielded dielectric waveguide antenna here we have two extra degrees of freedom provided by the geometry, to obtain the required beamwidth in the E- and in the H-plane, while keeping the launcher cross section.

The shielded solid dielectric horn is amenable to a simplified yet sufficiently accurate analysis, provided dielectric mouth cross section dimensions exceed about one wavelength. The radiation properties derived from the aperture fields using the Kirchhoff-Huygens vector formula are then used to build a set of design curves enabling the synthesis of such an antenna. Experimental evidence is provided in support of the design criteria.

The shielded solid dielectric horn is shown to perform essentially as a plane rectangular aperture with a quadratic phase error; as distinguished from the discussion in Lier [21], here the aperture fields are approximated by the transverse fields of a dielectric waveguide operated in the $TE_{1,1}$ to X (or the $mTE_{1,1}$ to X) mode.

4.2 FROM WAVEGUIDE TO HORN

Metallic sectoral horns can be conceived as rectangular metallic waveguides in which a pair of nonadjacent walls gradually tapers out. When the walls normal to the electric field are flared out, the sectoral horn is called an E-plane sectoral horn. Likewise an H-plane sectoral horn is obtained when the walls normal to the magnetic field are flared out. Figure 4.1 illustrates these concepts.

In turn, the metallic pyramidal horn may be seen as deriving from a rectangular metallic waveguide in which both walls gradually taper out. We will assume that at first the side walls normal to the electric field are flared out in the manner used to obtain an E-plane sectoral horn, followed by the side walls normal to the magnetic field (see Figure 4.2).

Compared with the open-ended metallic waveguide, metallic horns show improved impedance matching between the waveguide feed and free space. Simultaneously, since the aperture is increased in the E-plane, the H-plane, or both planes, beamwidth in the same plane decreases. Since the mouth is no longer an equiphase surface, as in the waveguide, mouth phase errors, which increase with horn length and flare angle, limit the directivity

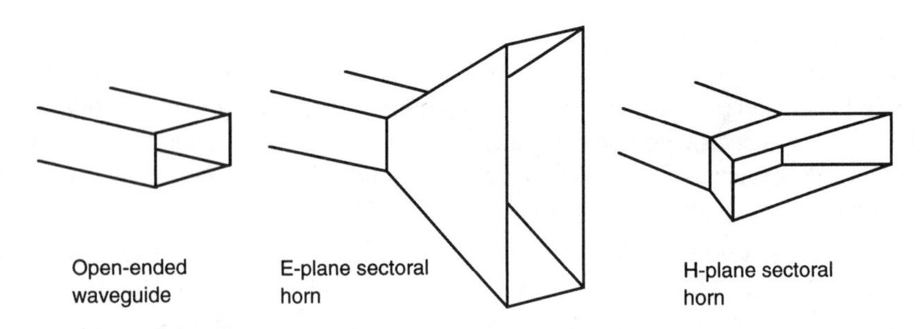

Figure 4.1: Metallic E- and H-plane sectoral horns derived from the open-ended metallic waveguide.

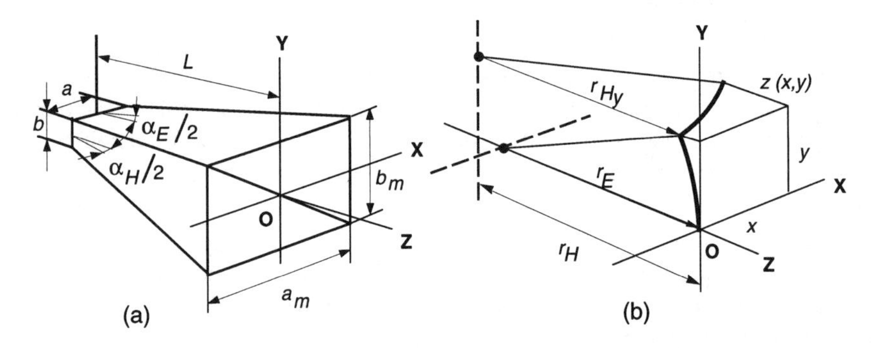

Figure 4.2: Metallic pyramidal horn derived from the open-ended metallic waveguide: (a) dimensions and (b) defining the phase at the mouth.

that can be achieved.

As the metallic horn transverse section increases there is a point where the $TE_{1,0}$ waveguide dominant mode is no longer the only propagating mode. To prevent these usually unwanted higher-order modes, the throat cross section should meet certain criteria [23] and the flare angle is often limited (to about 45 degrees).

The solid dielectric E-plane sectoral, H-plane sectoral, and pyramidal horns are henceforth referred to simply as the E-plane sectoral, the H-plane sectoral, and the pyramidal horns, and are the dielectric counterparts of the E-plane and the H-plane sectoral and the pyramidal metallic horns, respectively. Like the dielectric waveguide these solid dielectric horns may be fed by a rectangular metallic waveguide or by metallic sectoral or pyramidal horn operated in the $TE_{1,0}$ mode.

As in the metallic horn, flaring the dielectric waveguide provides the antenna designer with two extra degrees of freedom for shaping the radiation pattern. The underlying phenomena, however, are now far more complex than in the case of metallic horns. In the latter, flaring provided a means to increase mouth size (in one or both dimensions) while keeping the same type of waveguide field distribution, albeit for a mouth phase error. In the dielectric waveguide, besides the increase in mouth size and the introduction of a mouth phase error, we must consider the contributions from the side walls. In addition, the field structure changes along the horn length, due to the dependence of propagation constants with cross-section dimensions.

Shielded dielectric horns compare with shielded dielectric waveguide antennas very much in the same way as metallic horns and open-ended metallic waveguides and thus very simple design procedures, such as those described in Section 4.8, may be used.

4.3 E-PLANE SOLID DIELECTRIC SECTORAL HORN

The E-plane solid dielectric sectoral horn, shortly the E-plane sectoral horn is derived from a dielectric waveguide antenna in which the walls normal to the electric field (denoted top and bottom in Figure 4.4) flare out. As before we will assume that the launcher is a metallic rectangular waveguide operated in the $TE_{1,0}$ mode. Figure 4.3 shows an E-plane sectoral horn together with a coordinate system centered on the horn throat and defines the horn dimensions.

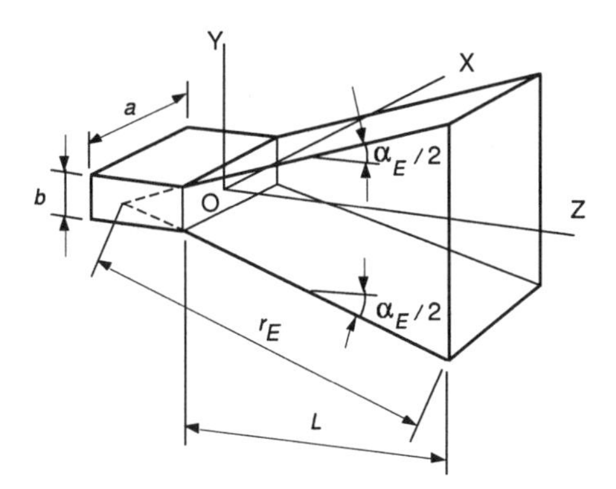

Figure 4.3: E-Plane sectoral horn and coordinate system.

Assume that, as far as the far-field radiation pattern is concerned, the horn behaves like a set of n dielectric waveguides (Figure 4.4) of length $l = L/n$, numbered from 1 to n from the horn throat to the horn mouth.

The cross-sectional dimensions of the ith waveguide, which will be a_i (along the X axis) and b_i (along the Y axis), are

$$a_i = a \tag{4.1}$$

$$b_i = b + 2(i-1)l\tan(\frac{\alpha_E}{2}) \tag{4.2}$$

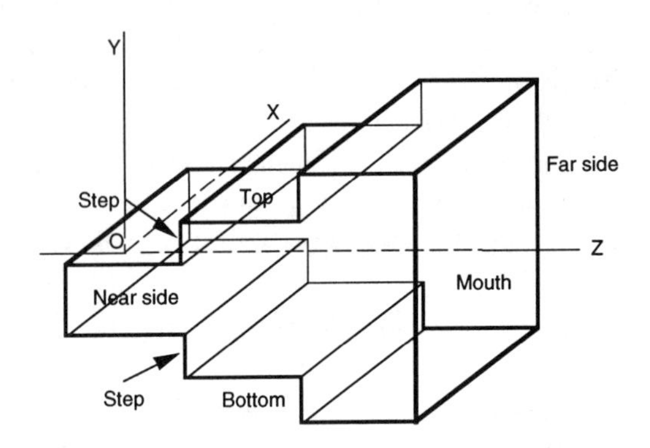

Figure 4.4: E-Plane sectoral horn as a set of gradually expanding waveguides.

The radiation pattern of the sectoral horn will be taken as an appropriate combination (weighed vector sum) of two radiation patterns:

- The sectoral horn alone, that is, ignoring the presence of the launcher;
- The launcher.

This is similar to the approach of James [22], except that the Kirchhoff-Huygens integration surface for the horn is taken as the whole dielectric surface rather than an infinite planar aperture at the horn mouth only. As for the launcher the measured pattern of the waveguide is considered rather than the result of the integration over a planar aperture on the horn throat.

We will start by the radiation pattern of the E-plane sectoral horn alone and we will calculate it by adding the following contributions, each being calculated using the standard vector Kirchhoff-Huygens formulas, presented in Chapter 2:

- Mouth (or free end) of the nth elementary waveguide;
- For elementary waveguides $i = 1 \cdots n$:
  - Top wall;
  - Bottom wall;
  - Near wall;
  - Far wall;
- For elementary waveguides $i = 2 \cdots n$:
  - Upper step;

– Lower step.

Taking an infinite number of elementary waveguides and considering the appropriate coupling between each, one would get an "exact" solution, that is, a solution that would be as good as the solution taken for the dielectric waveguide. In practice, if one takes into account a fairly small number of elementary waveguides and consider full transmission between adjacent elementary waveguides, this seems adequate in most cases, particularly when the flare angle is low to moderate.

When the launcher is a metallic waveguide operated in the $TE_{1,0}$ mode, the fields excited in the elementary waveguides may be TE to X or TM to Y. Since we only require the fields inside the waveguide there is no need to consider the mTE to X nor the mTM to Y modes. As we have shown (see Section 3.6) an X-band rectangular waveguide favors the TE to X mode. In the same way we will restrict our analysis to TE to X modes.

For most cases of practical interest, even when the first elementary waveguides can only support the $TE^X_{1,1}$ mode, as we progress toward the mouth and the cross section increases the $TE^X_{2,1}$, the $TE^X_{1,2}$ or even higher-order modes may reach propagating conditions. The presence of such modes may introduce significant changes in the overall radiation pattern due to the predominant role played here by the mouth. Since we are not attempting to include the effects of such modes, their presence must be minimized thus limiting the horn flare angle to low or moderate values (below about 45 degrees) and paying careful attention to the throat region and to the excitation.

Assuming that the propagating mode is the $TE^X_{1,1}$ in all elementary waveguides, the contributions from the top, bottom, near, and far walls may be calculated using (3.150), (3.151), (3.162), and (3.163), respectively, after substituting the field components by

$$E_{x_1} = 0 \tag{4.3}$$

$$E_{y_1} = j\beta \cos(k_{x_1}x)\cos(k_{y_1}y)[\exp(-j\beta z) + \Gamma\exp(+j\beta z)] \tag{4.4}$$

$$E_{z_1} = -k_{y_1}\cos(k_{x_1}x)\sin(k_{y_1}y)[\exp(-j\beta z) + \Gamma\exp(+j\beta z)] \tag{4.5}$$

$$H_{x_1} = +\frac{j\left(k_{x_1}^2 - k_1^2\right)}{\omega\mu_0}\cos(k_{x_1}x)\cos(k_{y_1}y)[\exp(-j\beta z) + \Gamma\exp(+j\beta z)] \tag{4.6}$$

$$H_{y_1} = -\frac{jk_{x_1}k_{y_1}}{\omega\mu_0}\sin(k_{x_1}x)\sin(k_{y_1}y)[\exp(-j\beta z) + \Gamma\exp(+j\beta z)] \tag{4.7}$$

$$H_{z_1} = +\frac{\beta k_{x_1}}{\omega\mu_0}\sin(k_{x_1}x)\cos(k_{y_1}y)[\exp(-j\beta z) + \Gamma\exp(+j\beta z)] \tag{4.8}$$

where the propagation constants, k_{x_1}, k_{y_1}, and β, inside the dielectric waveguides are in each case the solutions of the corresponding characteristic equation.

The complex voltage reflection coefficient Γ at $z = 0$ is given by

$$\Gamma = \frac{Z_0 - Z_1}{Z_0 + Z_1} \exp\left[-j2l\sum_{i=1}^{n}(\beta_i)\right] \tag{4.9}$$

where Z_1 is the transversal wave impedance at the mouth of the sectoral horn, Z_0 is free-space wave impedance, and β_i the propagation constant along the Z axis for the ith elementary waveguide

We must recall that besides the modifications of the field structure due to the varying cross section, field amplitudes also must be such that the power flow is constant along the Z axis. In addition, since the coordinate system for the waveguide was always set at the excitation interface, we have to include, in the one hand, the field delay due to mode propagation inside the sectoral horn, and on the other hand the usual factor $\exp[jk_0 l_i \cos(\theta)]$ where $l_i = (i-1)l$ is the distance along the Z axis between the origin of coordinates and the interface between the $i-1$ and the i elementary waveguides.

Assuming that the power transmitted along the Z axis is mostly confined to the sectoral horn cross section, an approximation which improves as the cross section increases, the contributions of the ith elementary waveguide should then be multiplied by

$$\sqrt{\frac{b_1}{b_i}} \exp[-jl(\sum_{m=1}^{i-1} \beta_m)] \exp[jk_0(i-1)l\cos(\theta)] \tag{4.10}$$

where the first term (square root) takes into account the variation in the cross section, the second term the propagation delay inside the horn, and the last term the displacement between the origin of coordinates for the horn and the origin of coordinates for each individual waveguide.

To compute the mouth contribution we cannot make direct use of the formulas derived for the dielectric waveguide because here we are confronted with a cylindric equiphase surface rather than with a plane one. For the usual sectoral horn configurations we may still use a plane mouth aperture provided we introduce a phase error to account for the wave propagation inside the sectoral horn.

Considering again Figure 4.3 it is easy to see that the phase error $\Delta\phi(x, y)$ at point x, y on the aperture is given by

$$\begin{aligned} \Delta\phi(x, y) &= -\beta(r_E - \sqrt{r_E^2 - y^2}) \\ &\approx -\beta\frac{y^2}{2r_E} \end{aligned} \tag{4.11}$$

As an example, this method was applied to an E-plane sectoral horn with $\bar{\varepsilon}_1 = 2.53$, $\alpha_E = 15$ degrees, $\bar{r}_E = 33.11$, and $\bar{L} = 25.5$, fed by a standard X-band metallic waveguide ($\bar{a} = 4.49$ and $\bar{b} = 1.99$) using $n = 20$ steps. The results obtained are shown in Figure 4.5.

In this figure the different components of the radiation pattern, mouth, top and bottom, near and far side walls and steps (defined in Figure 4.4) are represented separately thus showing clearly the predominance of the mouth radiation and the role of the side walls. The role of the steps is clearly very small and can safely be neglected.

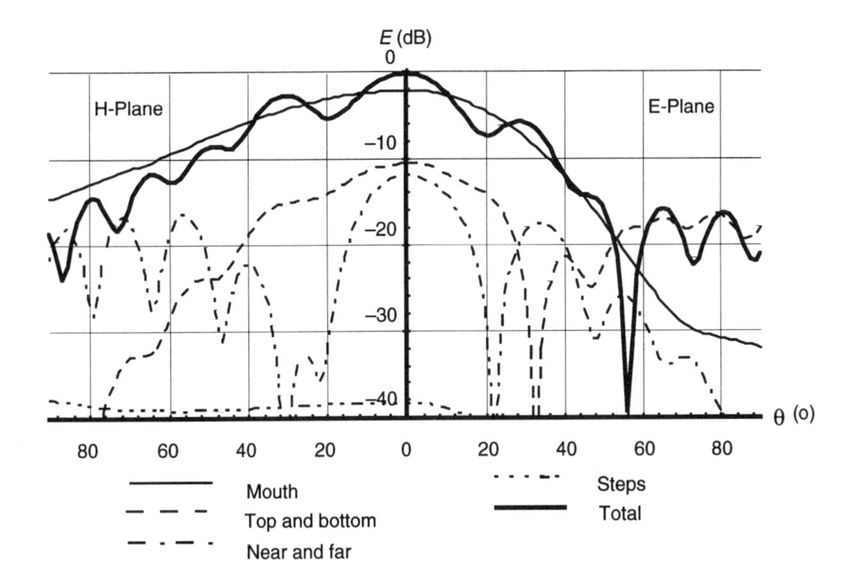

Figure 4.5: Computed E- and H-plane radiation patterns of an E-plane sectoral horn with $\bar{\varepsilon}_1 = 2.53$, $\alpha_E = 15$ degrees, $\bar{r}_E = 33.11$, and $\bar{L} = 25.5$, operated in the $TE^X_{1,1}$ mode, showing the contributions of the mouth, top and bottom, near and far walls, and steps.

In the manner explained in detail in Chapter 3, the launcher pattern was added to the computed horn pattern, taking $g_l = 4.64$ and $g_d = 13.58$, a launching efficiency of 0.9 and $k_p = 0.65$. The overall computed radiation patterns, normalized to boresight value, are shown in Figure 4.6 together with the measured patterns. The agreement between computed and measured patterns is excellent and extends well away from boresight, particularly in the H-plane.

The same method was successfully applied to other cases. A rather extreme one is also shown: an E-plane sectoral horn with $\bar{\varepsilon}_1 = 2.53$, $\alpha_E = 45$ degrees, $\bar{r}_E = 18.12$, and $\bar{L} = 16.5$, fed by a standard X-band metallic waveguide ($\bar{a} = 4.49$ and $\bar{b} = 1.99$) using $n = 10$ steps. The amplitude of the launcher pattern relative to the horn pattern was again calculated using (3.176) taking $g_l = 4.64$, $g_d = 21.24$, launching efficiency of 0.9, and $k_p = 0.78$. The computed and measured patterns, normalized to boresight and shown in Figure 4.7 agree very well, particularly in the H-plane.

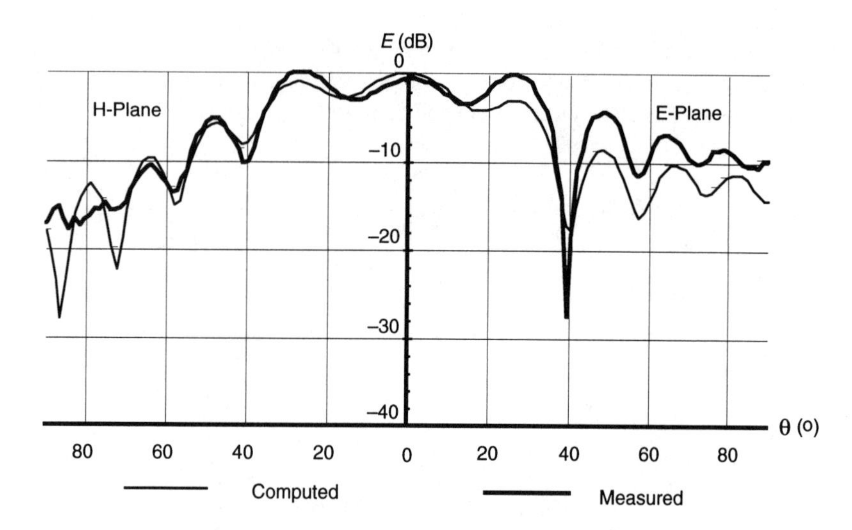

Figure 4.6: Computed and measured E- and H-plane radiation patterns of an E-plane sectoral horn with $\bar{\varepsilon}_1 = 2.53$, $\alpha_E = 15$ degrees, $\bar{r}_E = 33.11$, and $\bar{L} = 25.5$, fed by a standard X-band metallic waveguide ($\bar{a} = 4.49$ and $\bar{b} = 1.99$).

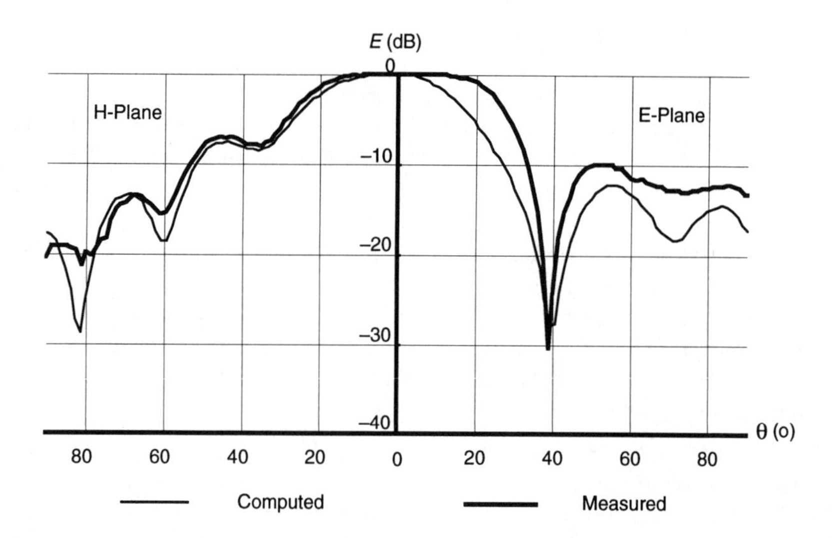

Figure 4.7: Computed and measured E- and H-plane radiation patterns of an E-plane sectoral horn with $\bar{\varepsilon}_1 = 2.53$, $\alpha_E = 45$ degrees, $\bar{r}_E = 18.12$, and $\bar{L} = 16.5$, fed by a standard X-band metallic waveguide ($\bar{a} = 4.49$ and $\bar{b} = 1.99$).

It should be stressed that the radiation patterns just presented are not very interesting for practical applications due to the high-sidelobe level. Nevertheless they provide valuable information on the radiation mechanisms that will be later used to turn solid dielectric horns into useful antennas and to derive a simplified design procedure.

It is instructive to examine the computed radiation patterns of the 45-degree E-plane sectoral horn alone, that is excluding the launcher, showing its various contributions (Figure 4.8) and then compare it to the 15-degree horn (Figure 4.5). In both cases the mouth plays the dominant role in the horn radiation pattern. As expected, since in both cases the mouth dimension in the H-plane is the same, the mouth radiation pattern in the H-plane is almost the same. In the E-plane the larger mouth size of the 45-degree horn, in spite of increased aperture phase error, leads to a sharper mouth diagram. The E-plane radiation pattern of the top and bottom walls shows a wider main lobe in the 45-degree horn due to flaring. The oscillatory nature of the radiation pattern of the near and far walls, particularly in the E-plane, is far more apparent in the 15-degree horn. This effect shows up quite clearly when comparing the E-plane patterns of both horns. Finally, the role of the steps although much higher in the 45-degree horn, is still minor.

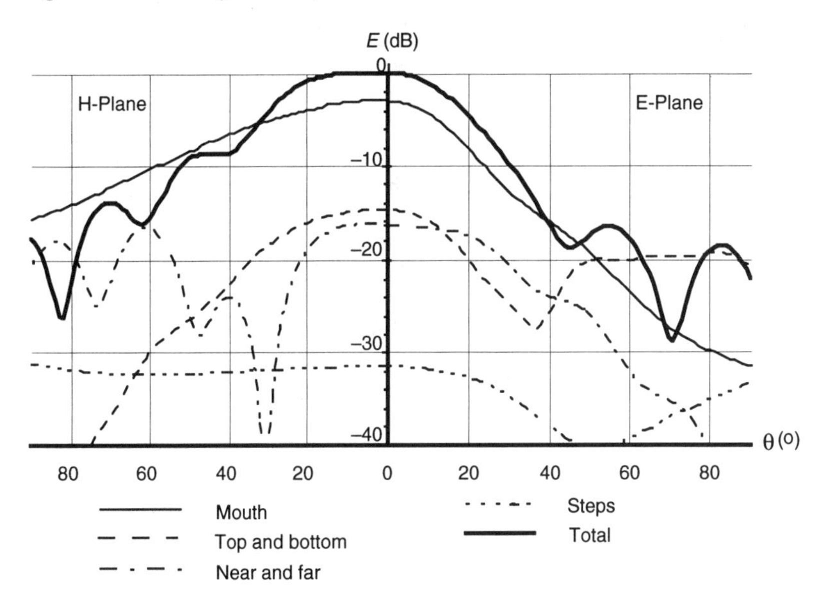

Figure 4.8: Computed E- and H-plane radiation patterns of an E-plane sectoral horn with $\bar{\varepsilon}_1 = 2.53$, $\alpha_E = 45$ degrees, $\bar{r}_E = 18.12$, and $\bar{L} = 16.5$, fed by a standard X-band metallic waveguide ($\bar{a} = 4.49$ and $\bar{b} = 1.99$), showing the contributions of the mouth, top and bottom, and near and far walls.

The design of an E- plane dielectric sectoral horn to achieve a given radiation pattern is a rather complex process about which only very general remarks can be made because of the large number of intervening factors: mouth, top and bottom walls, near and far walls, and launcher.

The mouth is usually the dominant factor. Gradual tapering of the field amplitude and phase error provide a nearly sidelobe-free radiation pattern which in the E-plane sectoral horn usually has a wider main lobe in the H-plane than in the E-plane. The mouth dimensions can be easily estimated from the desired radiation pattern, the major problem being that the mouth cannot be achieved without sidewalls. The mouth interaction with the side walls, in the one hand, and with the launcher in the other, modifies the mouth pattern introducing sidelobes. In some cases, as with the 45-degree horn, these interactions may produce a very flat main lobe, which may be desirable for some applications.

One possible way to reinforce the importance of the mouth in relation to the side walls is to make use of another launcher such that it keeps the former while reducing the latter. A metallic E-plane sectoral horn, with the same flare angle of the dielectric horn, is an obvious choice in that it preserves the mouth pattern while at the same time reducing the size and the role of the side walls. Figure 4.9 shows the geometry of an E-plane dielectric horn excited by an E-plane metallic sectoral horn with the same flare angle, where ul is the distance between the launcher mouth and the horn mouth.

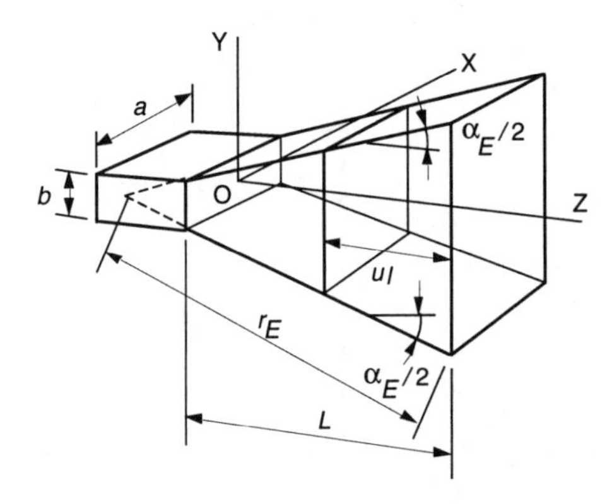

Figure 4.9: E-plane dielectric sectoral horn excited by an E-plane metallic sectoral horn.

Given the similarity between the modal fields of the $TE_{1,0}$ mode of a rectangular metallic waveguide and the $TE_{1,0}$ mode of a E-plane sectoral horn the launching efficiency of the $TE^X_{1,1}$ mode in the sectoral horn may be estimated from the data given in Section 3.6. Values of about 90% are easily achieved.

An extreme case, where ul is reduced to about 0.62λ, $\bar{ul} = 3.9$, is useful to illustrate the use of a metallic sectoral horn as a launcher for the E-plane sectoral horn, with $\bar{\varepsilon}_1 = 2.53$, $\alpha_E = 45$ degrees, and $\bar{L} = 16.5$. The calculated radiation pattern is given in Figure 4.10

where the calculated patterns include the contributions of the mouth, the top, the bottom, the near and the far sidewalls and neglects the steps. The computed (including the launcher effects) and measured radiation patterns are represented in Figure 4.11. The amplitude of the launcher pattern relative to the horn pattern was calculated as before taking $g_l = 9.55$ and $g_d = 8.86$, a launching efficiency of 0.9 and $k_p = 1$. The launcher pattern was calculated using the Kirchhoff-Huygens formulas.

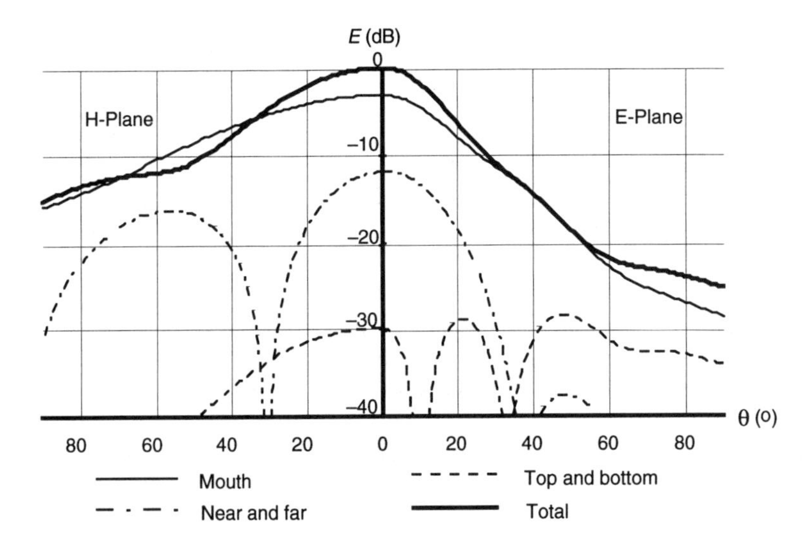

Figure 4.10: Computed far-field radiation patterns of an E-plane dielectric sectoral horn with $\bar{\varepsilon}_1 = 2.53$, $\alpha_E = 45$ degrees, $\bar{r}_E = 18.12$, and $\bar{L} = 16.5$ excited by a sectoral horn with the same flare angle and $\bar{ul} = 3.9$.

Comparing the radiation patterns when the launcher is a sectoral horn (Figure 4.11) and alternatively when it is a rectangular waveguide (Figure 4.7) we note that in the former the mouth pattern is more apparent and the importance of the side walls, particularly the near and far walls, is reduced.

Using such a long sectoral horn as the launcher is as far as we can go to reduce the effects of the side walls. Lower values of ul are likely to introduce significant coupling between the launcher aperture and the dielectric sectoral horn mouth, a factor that was neglected in our analysis. A much better solution will be dealt with in Section 4.7.

4.4 H-PLANE SOLID DIELECTRIC SECTORAL HORN

The H-plane solid dielectric sectoral horn, or simply the H-plane sectoral horn, is a dielectric waveguide antenna in which the walls normal to the magnetic field (also called

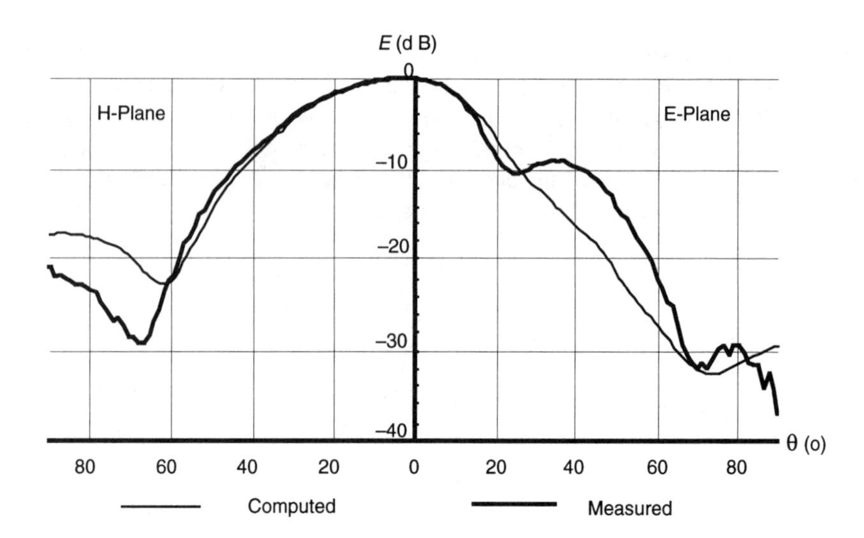

Figure 4.11: Computed and measured far-field radiation patterns of a E-plane sectoral horn with $\bar{\varepsilon}_1 = 2.53$, $\alpha_E = 45$ degrees, $\bar{r}_E = 18.12$, and $\bar{L} = 16.5$ using as the launcher an E-plane metallic sectoral horn with $\bar{ul} = 3.9$.

magnetic walls and in the following near and far walls) flare out. The launcher is assumed to be a metallic rectangular waveguide operated in the $TE_{1,0}$ mode.

As for the E-plane (solid dielectric) sectoral horn, to compute its radiation pattern, the H-plane sectoral horn will be modeled as a series of n gradually expanding dielectric waveguides of length $l = L/n$, numbered 1 to n from the throat to the mouth. The cross section dimensions of the ith waveguide are now

$$a_i = a + 2(i-1)l\tan(\frac{\alpha_H}{2}) \quad (4.12)$$

$$b_i = b \quad (4.13)$$

and the phase error at the aperture is

$$\Delta\phi(x, y) = -\beta(r_H - \sqrt{r_H^2 - x^2}) \approx -\beta\frac{x^2}{2r_H} \quad (4.14)$$

Proceeding just as for the E-plane sectoral horn, we computed the far-field radiation pattern of an H-plane sectoral horn with $\bar{\varepsilon}_1 = 2.53$, $\alpha_H = 20$ degrees, $\bar{r}_H = 38.7$, and $\bar{L} = 25.3$ excited by a standard X-band rectangular waveguide, taking $g_l = 4.64$, $g_d = 26.94$, excitation efficiency of 0.9, and $k_p = 0.65$. The computed and the measured patterns shown in Figure 4.12 are in good agreement.

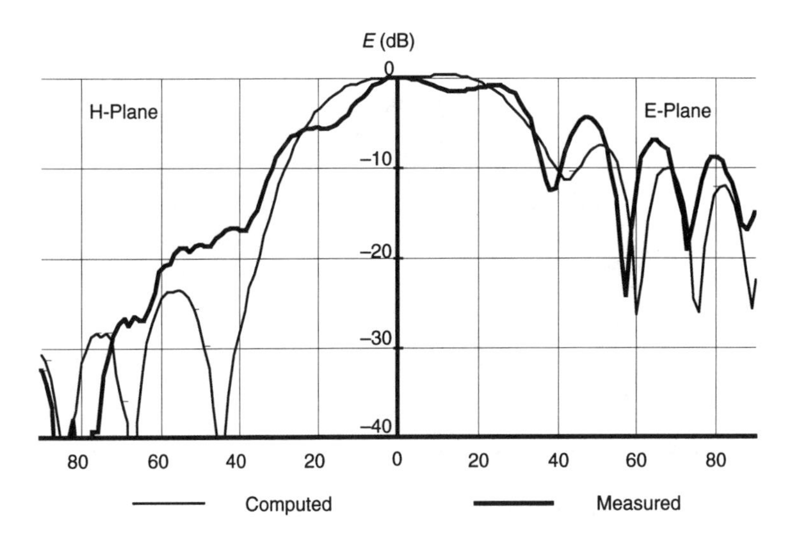

Figure 4.12: Computed and measured far-field radiation patterns of an H-plane sectoral horn with $\bar{\varepsilon}_1 = 2.53$, $\alpha_H = 20$ degrees, $\bar{r}_H = 38.7$, and $\bar{L} = 25.3$, fed by a standard X-band metallic waveguide ($\bar{a} = 4.49$ and $\bar{b} = 1.99$).

Figure 4.13 shows yet another example of computed and measured radiation patterns, now for the more extreme case of an H-plane sectoral horn with $\bar{\varepsilon}_1 = 2.53$, $\alpha_H = 45$ degrees, $\bar{r}_H = 20.6$, and $\bar{L} = 13.6$ excited by a standard X-band rectangular metallic waveguide and taking $g_l = 4.64$, $g_d = 21.09$, an excitation efficiency of 0.9 and $k_p = 0.65$. The agreement between computed and measured results deteriorates but is still acceptable.

The comments made in Section 4.3 about the usefulness of solid dielectric E-plane sectoral horns made also apply here.

Representing the various components of the sectoral horn computed radiation pattern (Figure 4.14) helps to explain the results. The mouth remains all-important in the overall radiation pattern and the top and bottom walls play a much more important role than in the E-plane sectoral horn (Figure 4.8). The steps have an even less significant role than in the E-plane sectoral horn which is due to the fact the field intensity on them in this case is far lower than in the previous one.

4.5 SOLID DIELECTRIC PYRAMIDAL HORN

The solid dielectric pyramidal horn, here simply the pyramidal horn, is the dielectric counterpart of the metallic pyramidal horn. It may be derived from a solid dielectric waveguide in which the top, bottom, and side walls flare out. The launcher is taken to be a metallic rectangular waveguide operated in the $TE_{1,0}$ mode, although other launchers

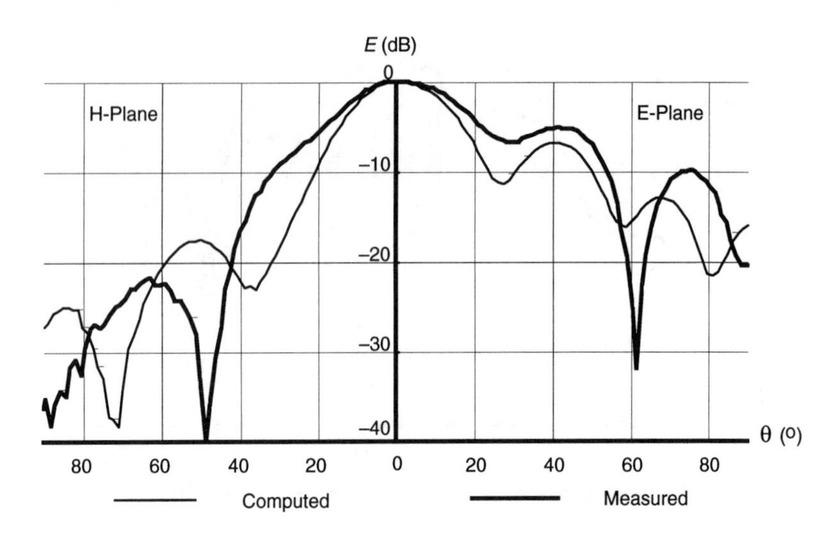

Figure 4.13: Computed and measured far-field radiation patterns of an H-plane sectoral horn with $\bar{\varepsilon}_1 = 2.53$, $\alpha_H = 45$ degrees, $\bar{r}_H = 20.6$, and $\bar{L} = 13.6$, fed by a standard X-band metallic waveguide ($\bar{a} = 4.49$ and $\bar{b} = 1.99$).

such as the metallic pyramidal horn may also be used.

Although flaring the dielectric waveguide to turn it into a pyramidal horn does offer two extra degrees of freedom for beam shaping, the associated design complexities are far more than in its metal counterpart. In the metal horn, flaring increases the aperture size and introduces a phase error but the field distribution remains essentially unchanged. In the solid dielectric horn, however, apart from the increase in mouth size and the introduction of a mouth phase error, the field structure changes along the horn length. Also, the contributions from the side walls cannot be neglected as in the metal horn.

To compute the far-field radiation pattern of the solid dielectric pyramidal horn we will use the method described in detail for the solid dielectric E-plane sectoral horn, assuming that the horn behaves like a set of n dielectric waveguides (see Figure 4.15) each of length $l = L/n$, numbered 1 to n from the horn throat to the horn mouth where

$$a_i = a + 2(i-1)l\tan(\frac{\alpha_H}{2}) \tag{4.15}$$

$$b_i = b + 2(i-1)l\tan(\frac{\alpha_E}{2}) \tag{4.16}$$

Assuming constant power flow the contributions of the ith elementary waveguide should be multiplied by

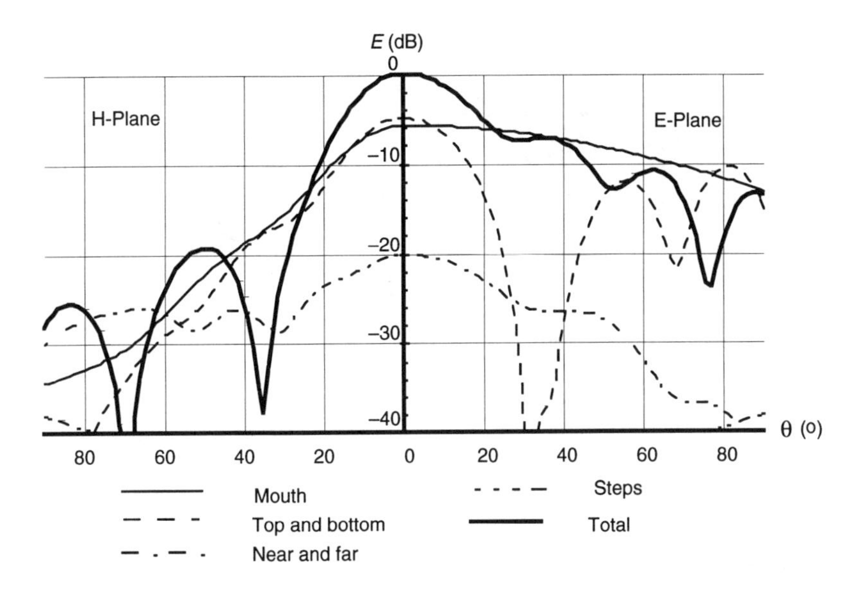

Figure 4.14: Computed far-field radiation patterns of an H-plane sectoral horn with $\bar{\varepsilon}_1 = 2.53$, $\alpha_H = 45$ degrees, $\bar{r}_H = 20.6$, and $\bar{L} = 13.6$, showing the contributions of the mouth, top and bottom, near and far walls, and steps.

$$\sqrt{\frac{a_1 b_1}{a_i b_i}} \exp[-jl(\sum_{m=1}^{i-1} \beta_m)] \exp[jk_0(i-1)l\cos(\theta)] \tag{4.17}$$

As the horn section increases toward the mouth, higher-order modes such as the $TE^X_{2,1}$ or the $TE^X_{1,2}$ may propagate and make their presence felt in spite of the fact that the first elementary waveguides support only the dominant $TE^X_{1,1}$ mode. To prevent the unwanted effects of these higher-order modes, which will be otherwise ignored, we confine ourselves to low or moderate flare angles.

For the usual pyramidal horn configurations we may still use a plane mouth aperture provided we introduce a phase error $\Delta\phi(x, y)$ to account for the wave propagation inside the horn.

Following our derivation of the pyramidal horn the equiphase surface on the aperture $z(x, y)$ may be taken as a curved surface such that its intersection with a set of planes, parallel to YOZ, is a set of arcs whose centers lie in a line parallel to the X axis at $z = -r_E$. Similarly, the intersection of the equiphase surface by a set of planes parallel to XOZ is a set of arcs whose centers lie in a line parallel to the Y axis at $z = -r_H$. For a general point, with coordinates x, y, on the equiphase surface we have

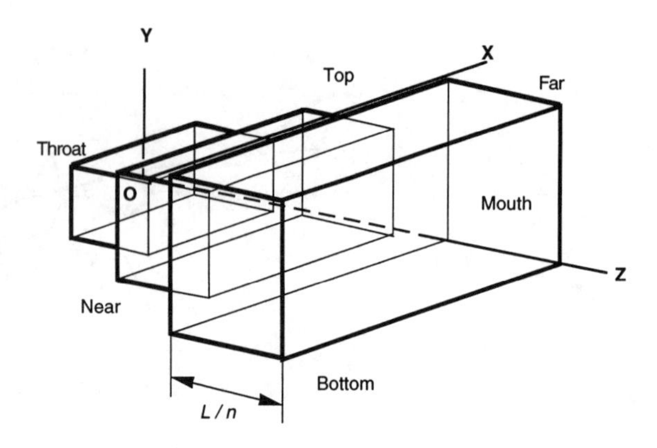

Figure 4.15: Pyramidal horn as a set of gradually expanding waveguides.

$$z(x, y) = -(r_E - \sqrt{r_E^2 - y^2}) - (r_{Hy} - \sqrt{r_{Hy}^2 - x^2}) \tag{4.18}$$

where

$$r_{Hy} = r_H - (r_E - \sqrt{r_E^2 - y^2}) \tag{4.19}$$

The phase error $\Delta\phi(x, y)$ at point (x, y) on the aperture is thus

$$\Delta\phi(x, y) = k_0 \bar{\beta}_n z(x, y) \tag{4.20}$$

Taking, for simplicity, $\frac{x}{2} \ll r_H$ and $\frac{y}{2} \ll r_E$ the aperture phase error becomes

$$\Delta\phi(x, y) \approx -k_0 \bar{\beta}_n \frac{x^2}{2r_H} - k_0 \bar{\beta}_n \frac{y^2}{2r_E} \tag{4.21}$$

As an example this method was applied to a pyramidal horn with $\bar{\varepsilon}_1 = 2.53$, $\alpha_E = 28$ degrees, $\alpha_H = 21$ degrees, $\bar{r}_E = 24.4$, $\bar{r}_H = 32.2$, and $\bar{L} = 19.6$, operated in the $TE_{1,1}^X$ mode and excited by a standard X-band rectangular metallic waveguide ($\bar{a} = 4.49$ and $\bar{b} = 1.99$) using $n = 10$ steps. The computed pattern indicating the various contributions is shown in Figure 4.16.

The launcher effect was calculated in the manner explained in detail in Section 3.8 taking $g_l = 4.64$, $g_d = 40.56$, an excitation efficiency of 0.9, and $k_p = 0.7$. The computed patterns, including the launcher effect, and the measured patterns are shown in Figure 4.17. Agreement between computed and measured patterns is fairly good.

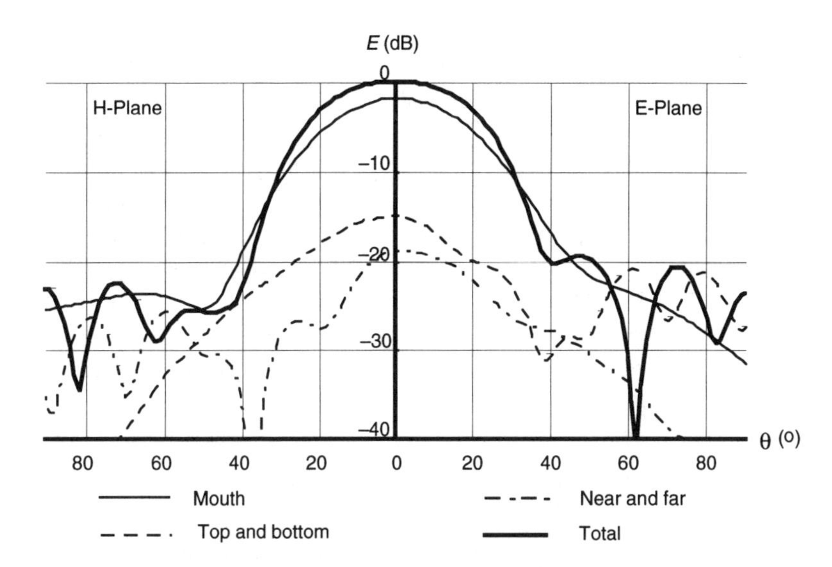

Figure 4.16: Computed far-field radiation patterns of a pyramidal horn with $\bar{\varepsilon}_1 = 2.53$, $\alpha_E = 28$ degrees, $\alpha_H = 21$ degrees, $\bar{r}_E = 24.4$, $\bar{r}_H = 32.2$, and $\bar{L} = 19.6$, fed by a standard X-band metallic waveguide ($\bar{a} = 4.49$ and $\bar{b} = 1.99$).

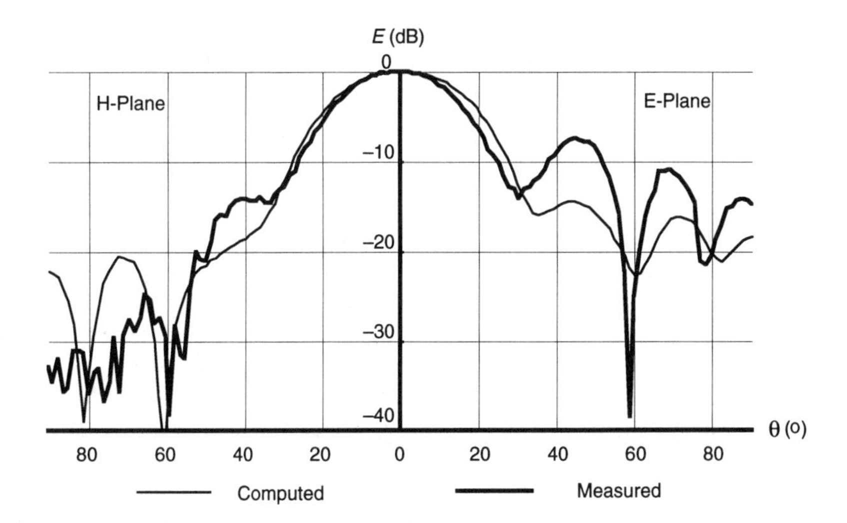

Figure 4.17: Computed and measured far-field radiation patterns of a pyramidal horn with $\bar{\varepsilon}_1 = 2.53$, $\alpha_E = 28$ degrees, $\alpha_H = 21$ degrees, $\bar{r}_E = 24.4$, $\bar{r}_H = 32.2$, and $\bar{L} = 19.6$, fed by a standard X-band metallic waveguide ($\bar{a} = 4.49$ and $\bar{b} = 1.99$).

As in the sectoral horn radiation pattern, here the mouth also plays the dominant role. With a square mouth the radiation patterns in both the E- and the H-plane are almost identical, the main disturbances being due to the launcher.

The remarks about the design of sectoral dielectric horn are also applicable here. The mouth is usually the dominant factor. The main lobe width is controlled by the aperture dimensions where gradual field amplitude tapering and phase error provide a low sidelobe radiation pattern. Once the aperture dimensions have been set the flare angles are established taking into account the launcher dimensions and whenever possible trying to minimize effects of the side walls.

When one is striving to achieve a circular symmetric pattern a very simple launcher, such as a rectangular waveguide which has markedly different beamwidths in the E- and the H-planes, may become a serious handicap.

4.6 SOLID DIELECTRIC HORNS' EXCITATION EFFICIENCY

The excitation efficiency of solid dielectric horns fed by a rectangular metallic waveguide may be computed in a manner akin to the one used for the dielectric waveguide. Here the only difference is that on the horn throat the horn fields have a phase error, given by (4.21), where for the E-plane sectoral horn $r_H = \infty$ and for the H-plane sectoral horn $r_E = \infty$. For the usual metallic waveguide dimensions and horn permittivities maximum phase variation is small (less than 1 radian) and the associated efficiency reduction can be neglected. Thus it is possible to estimate horn launching efficiency from the dielectric waveguide data supplied in Chapter 3.

4.7 SHIELDED SOLID DIELECTRIC HORNS

Just as in the case of the dielectric waveguide antenna, inserting a solid dielectric horn into a suitably sized metallic shield leads to a much improved radiation pattern with low or very low sidelobe levels. The shield shape is not critical in the resulting radiation pattern provided that, in the last wavelength or two near the aperture, its cross-sectional dimensions are such that the dielectric mode fields are negligible (< -20 dB referred to boresight) on the shield. Under these conditions the radiation pattern may be calculated using the Kirchhoff-Huygens formulas over the aperture, delimited by the shield. Unlike the shielded dielectric waveguide antenna the shielded dielectric horn (plane) aperture is no longer equiphase. The phase error follows a quadratic law with x, y as given by (4.11), (4.14), or (4.21) as appropriate. The aperture fields are simply obtained by adding a forward wave (moving from the excitation to the mouth) and a reflected wave traveling in the opposite direction.

For dielectric cross-section dimensions larger than about 1.5λ the TE to X fields may be used. For smaller horn sizes the mTE to X fields and the use of the "effective dielectric constant" method are required to obtain accurate radiation pattern predictions.

The shielded dielectric horn may be simply characterized by the relative dielectric permittivity $\bar{\varepsilon}_1$, dielectric and shield cross-section dimensions at the mouth, respectively a, b and a_m, b_m, and two parameters to define the phase error on the aperture. These parameters may be chosen as the phase errors Φ_E at $y = \pm b/2$ and Φ_H at $x = \pm a/2$. For the E-plane shielded dielectric sectoral horn $\Phi_H = 0$ while for the H-plane shielded dielectric sectoral horn $\Phi_E = 0$.

Figure 4.18 shows the computed and measured radiation patterns of a shielded solid dielectric E-plane sectoral horn with $\bar{\varepsilon}_1 = 2.53$, $\bar{a}_m = 13.19$, $\bar{b}_m = 9.70$, $\bar{a} = 4.49$, $\bar{b} = 3.76$, and $\Phi_E = 0.0525$ radians.

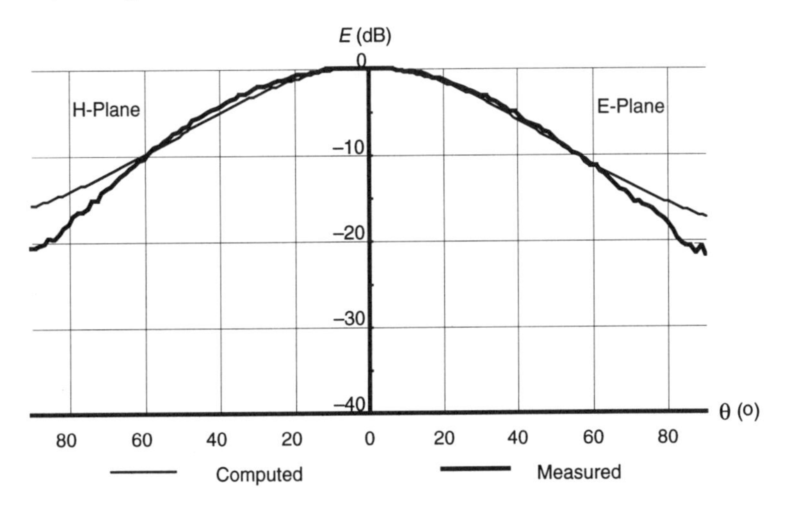

Figure 4.18: Computed and measured far-field radiation patterns of a shielded solid dielectric E-plane sectoral horn with $\bar{\varepsilon}_1 = 2.53$, $\bar{a}_m = 13.19$, $\bar{b}_m = 9.70$, $\bar{a} = 4.49$, $\bar{b} = 3.76$, and $\Phi_E = 0.0525$ radians.

Figure 4.19 shows the same data for a shielded solid dielectric H-plane sectoral horn with $\bar{\varepsilon}_1 = 2.53$, $\bar{a}_m = 14.86$, $\bar{b}_m = 14.92$, $\bar{a} = 13.31$, $\bar{b} = 1.99$, and $\Phi_H = 0.693$ radians. In both cases there is a remarkable agreement between computed and measured patterns.

Finally, Figure 4.20 shows the computed and measured radiation pattern of a shielded solid dielectric pyramidal horn with $\bar{\varepsilon}_1 = 2.53$, $\bar{a} = 8.82$, $\bar{b} = 4.12$, $\bar{a}_m = 13.19$, $\bar{b}_m = 9.70$, and $\bar{r}_E = \bar{r}_H = 49.50$. Also shown are the values obtained using the -3 dB and -10 dB design curves referred to in the next section.

From Figures 4.18 through 4.20 we may conclude that the shielded solid dielectric horn has much lower sidelobe levels than the corresponding solid dielectric horn. Also, since

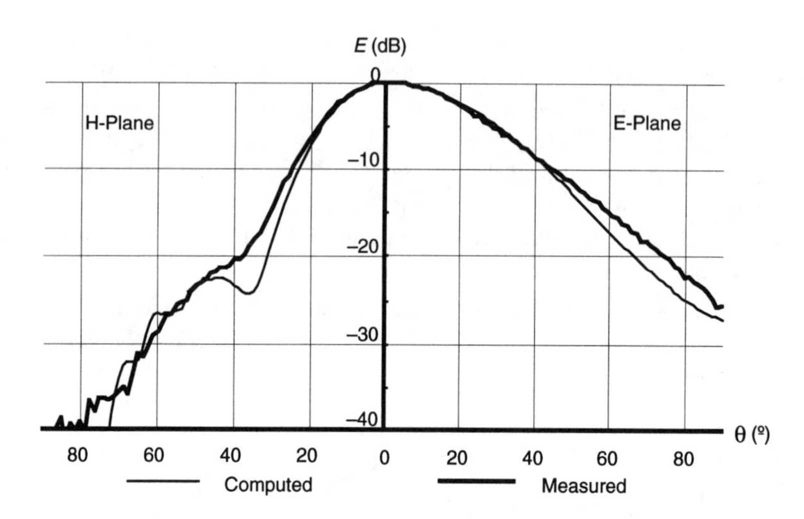

Figure 4.19: Computed far-field radiation patterns of a shielded solid dielectric H-plane sectoral horn with $\bar{\varepsilon}_1 = 2.53$, $\bar{a}_m = 14.86$, $\bar{b}_m = 14.92$, $\bar{a} = 13.31$, $\bar{b} = 1.99$, and $\Phi_H = 0.693$ radians.

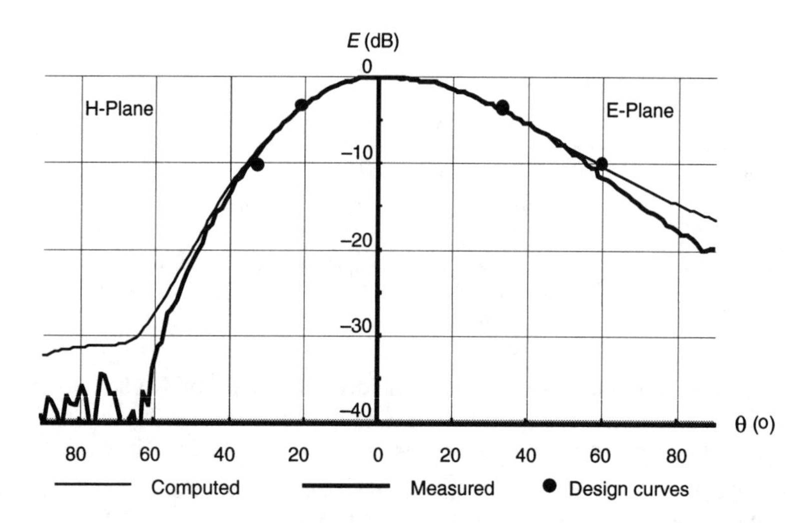

Figure 4.20: Computed and measured E- and H-plane radiation patterns for the $TE^X_{1,1}$ mode of a shielded solid pyramidal dielectric horn with $\bar{\varepsilon}_1 = 2.53$, $\bar{a} = 8.82$, $\bar{b} = 4.12$, $\bar{a}_m = 13.19$, $\bar{b}_m = 9.70$, and $\bar{r}_E = \bar{r}_H = 49.50$. Also shown are the values obtained using the -3 dB and -10 dB design curves.

its radiation pattern is mostly due to a single aperture, the phase center is well defined (slightly behind the aperture). Here we should note that the shield dimensions for the H-plane dielectric horn in Figure 4.19 are not optimal and thus sidelobe levels are higher then achievable with this type of antennas.

For moderate to large dielectric cross sections, typically found in pyramidal horns, we may approximate the aperture fields and derive very simple methods to compute the maximum directivity and the main lobe beamwidth. Direct antenna synthesis becomes possible and the electric design of the antenna is much simplified. These methods will be dealt with in the next section.

4.8 DESIGN OF SHIELDED SOLID DIELECTRIC PYRAMIDAL HORNS

In the first step of the design procedure we will approximate the shielded solid dielectric pyramidal horn, provided $\bar{a}, \bar{b} \geq 4$, by a rectangular aperture with a cosine-cosine illumination and a quadratic phase error and dimensions $\bar{a}_{ef}$ and $\bar{b}_{ef}$ given by

$$\bar{a}_{ef} = \frac{\pi}{\bar{k}_{x_1}} \tag{4.22}$$

$$\bar{b}_{ef} = \frac{\pi}{\bar{k}_{y_1}} \tag{4.23}$$

where $\bar{k}_{x_1}$ and $\bar{k}_{y_1}$ are the normalized transversal propagation constants for the dielectric waveguide with normalized cross-section dimensions $\bar{a}$ and $\bar{b}$.

For the $TE^X_{1,1}$ mode in most cases the values of $\bar{k}_{x_1}$ and $\bar{k}_{y_1}$ may be approximated by

$$\bar{k}_{x_1} \approx \frac{\pi}{\bar{a}}\left(1 - \frac{1}{1+\frac{\bar{a}}{2}\sqrt{\bar{\varepsilon}_1 - 1}}\right) \tag{4.24}$$

$$\bar{k}_{y_1} \approx \frac{\pi}{\bar{b}}\left(1 - \frac{1}{1+\frac{\bar{b}}{2}\sqrt{\bar{\varepsilon}_1 - 1}}\right) \tag{4.25}$$

Maximum phase errors, at the equivalent aperture boundaries $x = \pm a_{ef}/2$ and $y = \pm b_{ef}/2$, are

$$\Phi_H = \frac{\bar{\beta}\bar{a}_{ef}^2}{8\bar{r}_H} \tag{4.26}$$

$$\Phi_E = \frac{\bar{\beta}\bar{b}_{ef}^2}{8\bar{r}_E} \tag{4.27}$$

where as usual

$$\bar{\beta} = \sqrt{\bar{\varepsilon}_1 - \bar{k}_{x_1}^2 - \bar{k}_{y_1}^2} \tag{4.28}$$

As showed in Chapter 2, for a rectangular aperture with cosine-cosine illumination and a quadratic phase error, boresight directivity in dB over an isotropic source, is given by

$$10 \log_{10} \left(\frac{64\, \bar{a}_{ef}\, \bar{b}_{ef}}{\pi^5} \right) - \Delta_{\Phi_E} - \Delta_{\Phi_H} \tag{4.29}$$

where Δ_{Φ_E} are Δ_{Φ_H} are given in Figure 2.9 as a function of the maximum phase errors Φ_E and Φ_H.

To calculate the shield dimensions a_m and b_m at the mouth we will assume that the dielectric mode fields should decay to -20 dB of their value at the dielectric edge. With this criterion the radiation pattern sidelobe level should be about -30 dB below boresight level or less. Since $\bar{k}_{x_0}$ and $\bar{k}_{y_0}$ both tend to $\sqrt{\bar{\varepsilon}_1 - 1}$ for medium to large values of $\bar{a}$ and $\bar{b}$ we get

$$\bar{a}_m \approx \bar{a} + \frac{4.6}{\sqrt{\bar{\varepsilon}_1 - 1}} \tag{4.30}$$

$$\bar{b}_m \approx \bar{b} + \frac{4.6}{\sqrt{\bar{\varepsilon}_1 - 1}} \tag{4.31}$$

Figure 2.10 may now be used to estimate the far-field radiation pattern of a shielded dielectric horn. When maximum phase errors are smaller than about $\pi/4$ radians the radiation pattern has a dip, at $u \approx 4.7$, that tends to disappear as the maximum phase error increases. This corresponds in the H-plane to $\theta_0 \approx 9.4/\bar{a}_{ef}$ and in the E-plane to $\theta_0 \approx 9.4/\bar{b}_{ef}$. Under the same conditions the first sidelobe occurs at $u \approx 5.9$, that is, $\theta_1 \approx 11.8/\bar{a}_{ef}$ on the H-plane and $\theta_1 \approx 11.8/\bar{b}_{ef}$ on the E-plane, and its level is approximately $-28 + 20 \log_{10}(1 + \cos\theta_1)$ dB below boresight.

Approximating the shielded dielectric horn by a rectangular aperture with a cosine-cosine illumination with a quadratic phase error produces accurate enough estimates for the boresight directivity. However, results concerning sidelobes, namely, the main sidelobe level, tend to be overestimated by as much as 10 dB, which may prove to be unacceptable in many cases. This is due to the fact that the aperture fields follow a cosine law inside the dielectric but decay exponentially outside, a feature that leads to much lower sidelobe levels, usually -30 dB or less below boresight. Accurate estimates of the sidelobe level are thus left for a second step in the shielded dielectric horn design after the main dimensions have been defined.

For the first step of the design procedure we may simply make use of the above data or, if more detail such as the mainlobe -3 dB or -10 dB beamwidth is required, we may employ design curves. These curves have the drawback of being permittivity-dependent, but since they are based on actual TE to X aperture fields, they lead to more accurate results. Examples of such curves for $\bar{\varepsilon}_1 = 2.53$ and a_m and b_m related to a and b through (4.30) and (4.31) are give in Figures 4.21 and 4.22, which provide -3 dB and -10 dB beamwidths (in degrees), respectively, as a function of the normalized dielectric horn mouth dimensions in the same plane, taking as a parameter the maximum aperture phase error, at the dielectric border, of 0, $\pi/8$, $\pi/4$, and $\pi/2$ radians.

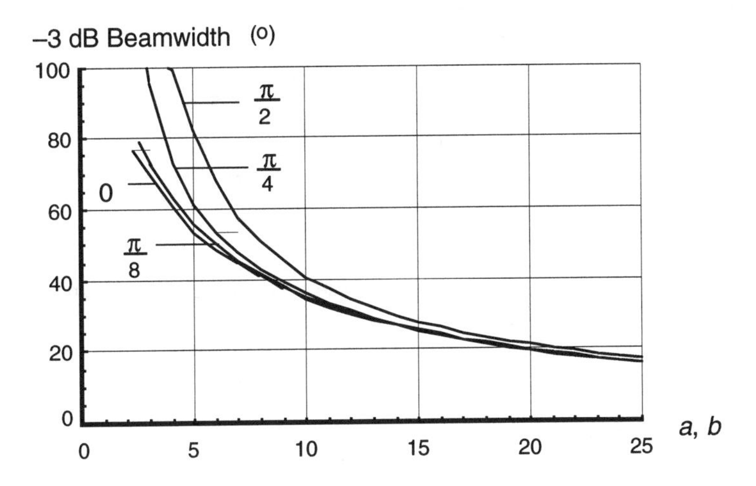

Figure 4.21: -3 dB beamwidth, in degrees, versus normalized dielectric dimension in the same plane, for $\bar{\varepsilon}_1 = 2.53$, taking as a parameter the maximum aperture phase error, in radians.

The beamwidths obtained using Figures 4.21 and 4.22 are plotted in Figure 4.20 for a shielded dielectric horn with $\bar{\varepsilon}_1 = 2.53$, $\bar{a} = 8.82$, $\bar{b} = 4.12$, $\bar{a}_m = 13.19$, $\bar{b}_m = 9.70$, and $\bar{r}_E = \bar{r}_H = 49.50$ which leads to maximum phase errors $\Phi_E = 0.124$ radians and $\Phi_H = 0.404$ radians. The agreement between computed, measured, and design values is very good.

Horn throat design is a subject still under active research and for which just general guidelines may be offered. Even if one usually wishes to launch only the $\mathrm{mTE}^X_{1,1}$ mode one cannot prevent the excitation of other higher-order modes whose presence in the radiation pattern results in higher cross-polarization, worse circular symmetry, higher sidelobe levels, and a more pronounced frequency dependence.

We will assume the launcher to be a rectangular metallic waveguide (eventually dielectric filled) operated in the $\mathrm{TE}_{1,0}$ mode. Following Figure 4.23 the metallic waveguide normalized cross-section dimensions $\bar{a}_g$ and $\bar{b}_g$ obey the following conditions for the $\mathrm{TE}_{1,0}$

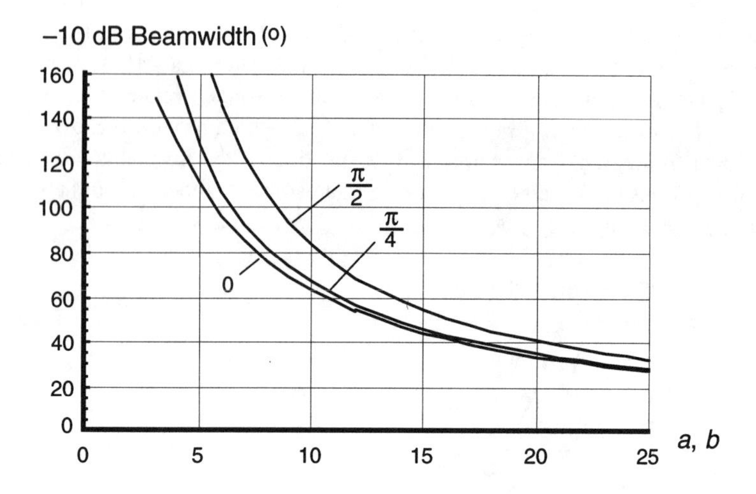

Figure 4.22: -10 dB beamwidth, in degrees, versus normalized dielectric dimension in the same plane, for $\bar{\varepsilon}_1 = 2.53$, taking as a parameter the maximum aperture phase error, in radians.

to be the only propagating mode:

$$2\pi > \bar{a}_g > \pi \tag{4.32}$$

$$\pi > \bar{b}_g > 0 \tag{4.33}$$

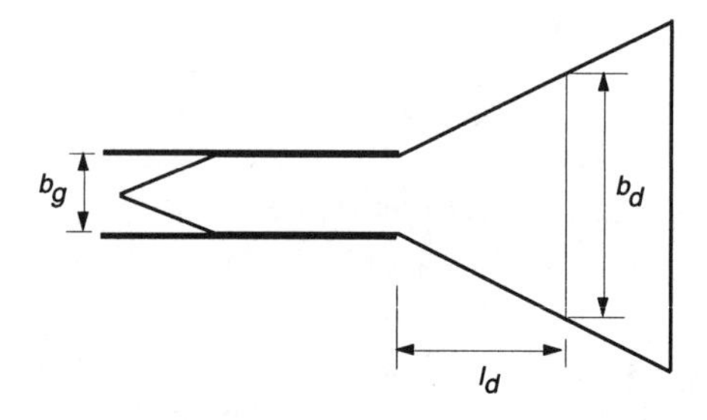

Figure 4.23: Dielectric horn throat region.

Taking, for example, the standard X-band rectangular waveguide at 9 GHz we have $\bar{a}_g = 4.31$ and $\bar{b}_g = 1.92$.

Extending the dielectric inside the metallic waveguide is equivalent to multiplying the cross dimension by $\sqrt{\bar{\varepsilon}_1}$. Taking the previous example with $\bar{\varepsilon}_1 = 2.53$ this corresponds to $\bar{a}_g = 6.85$ and $\bar{b}_g = 3.04$, which means that the $TE_{2,0}$ mode is now in the propagating region inside the dielectric-filled metallic waveguide. Since it is odd, the $TE_{2,0}$ mode of the metallic waveguide will not affect the excitation of even modes in the dielectric horn, however it will give rise to odd dielectric modes. If these modes are in the propagating region, they may distort the aperture fields and consequently the radiation pattern.

If the transition between air to dielectric inside the metallic waveguide is wedge shaped, normal to the $TE_{1,0}$ mode electric field lines, the $TE_{2,0}$ mode is not likely to be excited to any significant extent.

The next higher-order mode in the standard X-band metallic waveguide is the $TE_{1,1}$ mode for which

$$2\pi > \bar{a}_g > \pi \tag{4.34}$$

$$2\pi\sqrt{1+\left(\frac{9}{4}\right)^2} > \bar{b}_g > \pi\sqrt{1+\left(\frac{9}{4}\right)^2} \tag{4.35}$$

If $\bar{b}_g \geq 7.74$, which will happen for a standard X-band dielectric-filled ($\bar{\varepsilon}_1 = 2.53$) metallic waveguide for frequencies above 10.2 GHz, the $TE_{1,1}$ may propagate inside the metallic waveguide and will interfere with the excitation of the $TE^X_{1,1}$ dielectric mode but will not excite odd dielectric modes.

Looking into the dielectric horn, we find out that the first higher-order mode likely to be excited by the $TE_{1,0}$ mode of the metallic waveguide is the $mTE^X_{1,2}$; using the "effective dielectric constant method," the cutoff condition for this mode on a dielectric cross section with normalized dimensions $\bar{a}_d$ and $\bar{b}_d$ is

$$\bar{b}_d = \frac{2\pi}{\sqrt{\bar{\varepsilon}_1 - 1}} \tag{4.36}$$

Taking $\bar{\varepsilon}_1 = 2.53$ we get $b_d = 5.08$, which means that the $mTE^X_{1,2}$ is below cutoff at the throat when the launcher is a standard X-band metallic waveguide. The distance l_d between the waveguide end plane and the onset of this mode in the dielectric horn decreases with the flare angle (see Figure 4.23). Thus, for large flare angles the evanescent modes excited at the throat are more likely to reach the propagating region with a large enough amplitude. A value of $k_0 l_d \approx 2\pi$ appears to be an acceptable design compromise. For $\bar{\varepsilon}_1 = 2.53$ and $\bar{b}_g = 1.92$ the above criteria yields $\alpha_E/2 < 14.1$ degrees.

Shield dimensions near the throat should be such that the metal perturbs as little as possible the dielectric mode fields. In practice this means that the shield should be further away from the dielectric in the throat region than in the aperture.

4.9 CONCLUSIONS

The radiation pattern of low to moderate (up to 45 degrees) flare angle solid dielectric sectoral and pyramidal horns, excited by rectangular metallic waveguides or metallic horns, with the same flare angle as the dielectric horn, can be predicted, fairly accurately, adding up the independent contributions from the horn and the launcher. The latter is assumed to radiate unaffected by the presence of the horn except for a reduction in radiated power due to power being converted into horn modes and a shift in the position of the phase center due to the propagation delay inside the dielectric. The radiation pattern of the horn is calculated by assimilating it to a series of gradually expanding dielectric waveguides. Solving the characteristic equation for each of these waveguides and adjusting amplitudes so as to achieve power conservation yields the fields in each elementary waveguide. The radiation patterns are then derived from these fields using the vector Kirchhoff-Huygens formula. Agreement between computed and measured patterns validates the approach.

The design of solid dielectric horns is a difficult task due to the number, complexity, and interplay of the factors involved. In general terms we can say that the radiation pattern is dominated by the effect of the mouth, predicted with relative ease from the mouth dimensions and the mouth radius r_E and r_H which determine mouth phase error. Sidewalls and launcher radiation, even if often lower than the mouth, can adversely affect the radiation pattern particularly in the region of the far out sidelobes. However, careful dimensioning may result in an unusually flat main lobe and reasonably good circular symmetry (at least in the main lobe).

The importance of the sidewalls relative to the mouth may be reduced by using as the launcher a metallic horn (with the same flare angle as the dielectric horn). Given the similarity between the metallic horn and the metallic rectangular waveguide fields and since the flare angles in both the launcher and the dielectric horn are the same, the excitation efficiency may be computed just as in the dielectric waveguide case.

Inserting the solid dielectric horn into a suitably shaped and sized metallic shield leads to a dramatic improvement of the radiation characteristics. The shield effectively blocks the radiation from the launcher and the sidewalls; thus, the resulting radiation pattern, now due mainly to the horn mouth and surrounding area limited by the shield, shows much lower sidelobe levels and a well-defined phase center. In addition, since radiation mechanisms are much simpler, direct antenna synthesis becomes possible and electrical antenna design is that much easier.

Approximating the radiation pattern of the shielded solid dielectric horn by the radiation pattern of a rectangular aperture with a cosine-cosine illumination with a quadratic phase error provides a first and very general yet simple design procedure. Design curves, representing the -3 dB and -10 dB mainlobe beamwidths versus the dielectric horn mouth dimensions, enable a more accurate design.

Although here we do not deal with the cross-polarization properties of the shielded dielectric horn, these have been measured for a number of cases. Typical maximum cross-polarization levels are 30 and even 35 dB below copolar boresight level.

Impedance matching is another design parameter we do not discuss here. Using a number of standard gain X-band pyramidal horns as shields, measured reflected power levels, at the feed port, are between −15 and −20 dB over a large bandwidth (often the whole of X band).

The shielded solid dielectric horn has a number of interesting features: reasonably high aperture efficiency, low sidelobe level, low cross-polarization, good circular symmetry (when required), and reasonable impedance matching over a large bandwidth. In addition, it is relatively simple and inexpensive to manufacture when compared to other antennas with similar performances, such as corrugated horns, particularly for the millimeter and submillimeter bands. The penalty is a slightly bulkier antenna which may be acceptable, particularly at millimeter wavelength.

REFERENCES

[1] Mueller, G. E., and Tyrrell, W. A., (1947), Polyrod Antennas, *Bell Systems Technical Journal*, Vol. 26, pp. 837–857.

[2] Wilkes, G., (1948), Wavelength Lenses, *Proceedings of the Institute of Radio Engineers*, Vol. 36, pp. 206–212.

[3] Watson, R. B. and Horton, C. W. (1948), The Radiation Patterns of Dielectric Rods – Experiment and Theory, *Journal of Applied Physics*, Vol. 19, pp. 661–670.

[4] Shiau, Y., (1976), Dielectric Rod Antennas for millimeter Wave Integrated Circuits, *IEEE Microwave Theory and Techniques*, Vol. MTT-24, pp. 869–872.

[5] Kobayashi, S., Mittra, R., and Lampe, R., (1982), Dielectric Tapered Rod Antennas for Millimeter Wave Applications, *I.E.E.E. Transactions on Antennas and Propagation*, Vol. AP-30(1), pp. 54–58.

[6] Paulit, S. K., and Chatterjee, R., (1983), Radiation Characteristics of the Tapered Rectangular Dielectric Rod Antenna at Microwave Frequencies, *Proceedings of the 1983, URSI Symposium on Electromagnetic Theory, Santiago de Compostela*, Spain, pp. 401–404.

[7] Ohtera I., and Ujiie, H., (1981), Radiation Performance of a Modified Rhombic Dielectric Plate Antenna, *IEEE Transactions on Antennas and Propagation*, Vol. AP-29, No. 4, pp. 660–662.

[8] Jha, B., (1984), Sectoral Dielectric Horn Antennas, Ph.D. Thesis, Banaras Hindu University, India.

[9] Singh, A. K., (1991), *Study of Some Aspects of Sectoral Dielectric Horn Antennas*, Ph.D. Thesis, Banaras Hindu University, India.

[10] Jha, B., and Jha, R. K., (1984), Surface Fields of H-Plane Sectoral Dielectric Horn Antenna, *Proceedings of the 15th European Microwave Conference.*

[11] Singh, A. K., Jha, B. and Jha, R. K., (1991), Near-Field Analysis of E-Plane Sectoral Solid Dielectric Horn Antennas, *International Journal of Electronics*, Vol. 71, No. 4, pp. 697–706.

[12] Brooking, N., Clarricoats, P. J. B., and Olver, A. D., (1974), Radiation Pattern of Pyramidal Dielectric Waveguides, *Electronics Letters*, Vol. 10, pp. 33–34.

[13] Peace, G. M., and Swartz, E. E. (1964), Amplitude Compensated Horn Antenna, *The Microwave Journal*, February, pp. 66–68.

[14] Lawrie, R. E. and Peters L. (1966), Modifications of Horn Antennas for Low sidelobe Levels, *IEEE Transactions on Antennas and Propagation*, Vol. AP-14, No. 5, pp. 605–610.

[15] LaGrone, A. H. and Roberts, G. F. (1966), Minor Lobe Suppression in a Rectangular Horn Antenna Through the Utilization of a High Impedance Choke Flange, *IEEE Transactions on Antennas and Propagation*, Vol. AP-14, No. 1, pp. 102–104.

[16] Mentzer, C. A., and Peters, L., (1976), Pattern Analysis of Corrugated Horn Antennas, *IEEE Transactions on Antennas and Propagation*, Vol. AP-24, No. 3, pp. 304–309.

[17] Nair, R.A., Kamal, A. K., and Gupta, S. C., (1978), A High-gain Multimode Dielectric-coated Rectangular Horn Antenna, *The Radio and Electronic Engineer*, Vol. 48. No. 9, pp. 439–443.

[18] Chan, K. K., Huang, C. C., and Raab, A. R., (1980), Dielectric Loaded Trifurcated Horn for H-plane Stacked Reflector Feed Arrays, *Proceedings of the IEE*, Vol. 127(H), pp. 61–64.

[19] Sabnani, K. K. and Arora, R. K., (1978) Radiation Characteristics of a Rectangular Aperture Centrally Loaded with a Dielectric Slab in the H-plane, *Journal of the Institute of Electronic and Telecommunication Engineers (India)*, Vol. 24. No. 7, pp. 302–304.

[20] Marcatili, E. A. C., (1969), Dielectric Rectangular Wave Guide and Directional Coupler for Integrated Optics, *Bell System Technical Journal*, Vol. 48, No. 7, pp. 2071 – 2102.

[21] Lier, E., and Stoffels, C., (1991), Propagation and Radiation Characteristics of rectangular dielectric-loaded hybrid mode horn, *Proceedings of the IEE (H)*, Vol. 138, No. 5, pp. 407–411.

[22] James, J. R., (1967), Theoretical Investigations of Cylindrical Dielectric Rod Antennas, *Proceedings of the IEE* Vol. 114, pp. 309–319.

[23] Kraus, J. D., (1950), *Antennas*, McGraw-Hill.

Chapter 5

Dielectric Cylinder

5.1 INTRODUCTION

In her excellent review of dielectric cylindrical antennas,Chatterjee [1] cites Hondros and Debye [2] as the earliest work, in 1910, on the propagation of electromagnetic waves along dielectric cylinders.

The use of the dielectric cylinder as an antenna, also known as the dielectric rod antenna (Figure 5.1), has been known since 1939 [3] but, for a long time, the mechanisms involved in predicting the radiation patterns have been subject to a good deal of controversy. James [4] reviewed the existing theories, due to Horton [5], Bouix [6], Fradin [7], and Brown and Spector [8], and established that the patterns were the result of interference produced by the radiation coming from surfaces S_1 and S_2 (Figure 5.1).

Andersen [10] has shown that even if an infinite lossless dielectric cylinder carrying a propagating mode does not radiate, a finite cylinder due to the discontinuities at the launcher and the free end radiates, both from the free end and the side wall.

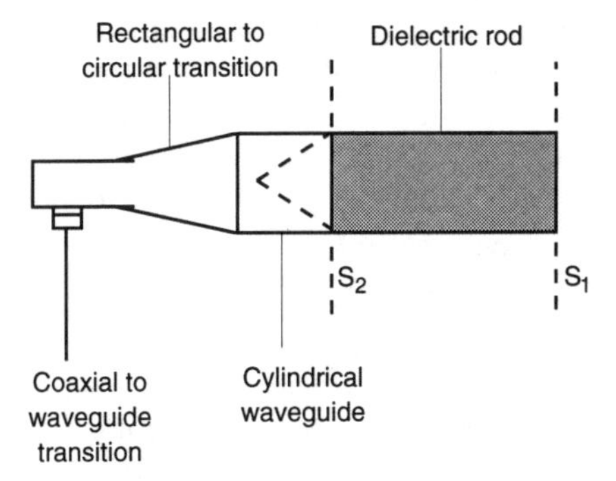

Figure 5.1: Dielectric rod antenna.

Newmann [9] established experimentally that, when the radiation from the launcher is effectively suppressed, the pattern of a very long (320 λ) dielectric rod can be accurately predicted assuming that the aperture field at S_1 is due only to a guided mode. In the general case, prediction is more difficult because, although the guided mode fields at S_1 can be accurately calculated, the same is not true for the fields associated with the contribution from the launcher end (S_2), where a continuous spectrum wave decaying away from S_2 exists. In many practical cases the effect of the launcher on the overall radiation pattern may be decisive, especially at the sidelobe level.

Yaghjian and Kornhauser [11] presented an approximate solution for the transition between a semi-infinite dielectric cylinder and free space and thus improved on the usual assumption of taking the unperturbed mode fields on S_1.

Using a curve fitting technique to obtain the excitation efficiency, and assuming that the radiation from the launcher was not modified by the presence of the dielectric rod, except for a constant amplitude factor (due to the energy converted into the modes), James [4] was able to predict the radiation pattern of dielectric rod antennas. His method works well except when the dielectric rod is very short and there is significant coupling between surfaces S_1 and S_2.

The transition between the launcher and the dielectric cylinder poses a complex problem. For the general case only approximate solutions are available mostly based on the modal matching technique described in Collin [12]. Rigorous solutions obtained by the Wiener-Hopf method were presented among others by Angulo and Chang [13], for the excitation efficiency of the transverse magnetic (TM) modes in the dielectric by the $TM_{0,1}$ mode of a metallic cylinder with the same radius.

The dielectric rod has been extensively researched and reviewed [1, 10] either as an antenna by itself or as a feed for reflector antennas. On the last topic the reader is referred to a recent and thorough survey by Olver [14].

Normally dielectric rod antennas make use of materials with a relative dielectric constant of the order of 2.5, have radius of the order of a half-wavelength or less and are operated in the $HE_{1,1}$ mode. As such they may be used as a simple inexpensive end-fire antenna with gains in the range 13 to 16 dBi and similar radiation patterns in both the E- and the H-planes [14], their main disadvantage being a high backlobe level (about -20 dB) due to the backward propagating wave reflected at the rod free end. Somewhat improved performance can be obtained by tapering the dielectric to reduce its radius towards the free end.

In 1966, Lewis [15] pointed out that with large-diameter, low-permittivity ($\bar{\varepsilon}_1 \approx 1.08$) dielectric foam cylinders, operated in the $HE_{1,1}$ hybrid mode and excited by pyramidal horns, very high gains ($\approx$30 dBi) could be obtained. The use of a low-permittivity dielectric ensures that only the $HE_{1,1}$ mode can propagate.

In general, the terms dielectric cylinder and dielectric rod antennas have been used as synonymous. Here we will use dielectric rod for dielectric cylinder antennas with relative dielectric permittivity around 2.5 and dielectric foam cylinder antennas for the case of very low values of the relative dielectric permittivity (< 1.2).

The similarity of behavior between a corrugated cylindrical waveguide and the dielectric rod, pointed by Clarricoats [16], suggests that this structure may provide an inexpensive alternative to the corrugated waveguide. The fact that the low flare angle cone can be approximated by a gradually expanding cylinder ensures that the study of the excitation and radiation properties of the cylinder provides a good insight into the behavior of the dielectric cone.

In this chapter we start by deriving a hybrid mode solution to Maxwell's equations for an infinite lossless dielectric cylinder in free space. After discussing rigorous and approximate solutions to the characteristic equation we proceed to describe, in some detail, the dominant $HE_{1,1}$ mode fields. The excitation efficiency of this mode by an arbitrary incident field is then calculated using modal matching techniques.

The radiation properties of the dielectric cylinder antenna operated in the $HE_{1,1}$ mode are discussed and a simple procedure, backed by experimental evidence, is suggested to predict the effect of the launcher.

Finally we approximate the electromagnetic fields of the $HE_{1,1}$ mode of a dielectric cylinder by a Gaussian beam and making use of the paraxial method of analysis derive a simple method of calculating the near- and far-field radiation patterns.

5.2 A SOLUTION OF MAXWELL'S EQUATIONS FOR A DIELECTRIC CYLINDER

Unlike all the other dielectric structures dealt with in this book, such as the rectangular cross section waveguide, the sectoral and the pyramidal horns and the cone, an exact solution of Maxwell's equations for the infinite dielectric cylinder exists [12, 17]. Due to its importance for the dielectric cone we will revise the subject in the following.

Consider an infinite isotropic lossless homogeneous dielectric cylinder of radius ρ_1 immersed in otherwise free space, as shown in Figure 5.2 and a cylindrical system of coordinates ρ, φ, z. We will be looking for solutions where the vector potentials $\mathcal{A}$ and $\mathcal{F}$ take the general form

$$\mathcal{A} = A_z \boldsymbol{z} \quad (5.1)$$

$$\mathcal{F} = F_z \boldsymbol{z} \quad (5.2)$$

where $\boldsymbol{z}$ is the unit vector in the direction of the Z axis.

Assuming an harmonic time dependence and dropping a common $\exp(j\omega t)$ factor, A_z and F_z will take the form

$$A_{z_1} = a_c J_n(k_{\rho_1}\rho)\sin(n\varphi)\exp(\pm j\beta z) \quad (5.3)$$

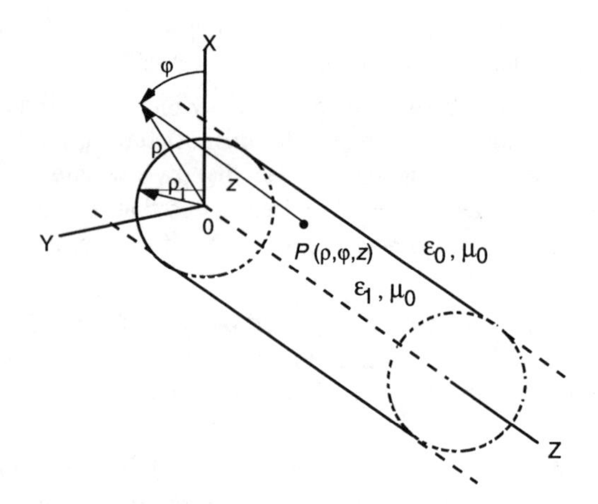

Figure 5.2: Coordinate system for the dielectric cylinder.

$$A_{z0} = b_c K_n(k_{\rho 0}\rho)\sin(n\varphi)\exp(\pm j\beta z) \tag{5.4}$$
$$F_{z1} = c_c J_n(k_{\rho 1}\rho)\cos(n\varphi)\exp(\pm j\beta z) \tag{5.5}$$
$$F_{z0} = d_c K_n(k_{\rho 0}\rho)\cos(n\varphi)\exp(\pm j\beta z) \tag{5.6}$$

where $J_n(x)$ and $K_n(x)$, respectively, denote the Bessel function of first kind and the modified Bessel function of second kind, order n and argument x, β is the longitudinal propagation constant, k_{ρ_1} and k_{ρ_0} are the radial propagation constants, and the subscripts 1 and 0 denote the regions inside and outside the cylinder, respectively. Constants a_c,b_c, c_c, and d_c are to be determined by matching the boundary conditions. In the above expressions the minus sign corresponds to a wave traveling in the positive direction of the Z axis and the plus sign to a wave traveling in the opposite direction.

We could have taken the cosine function for A_z and the sine function for F_z. In that case we would end up with a similar set of fields except for the direction of polarization which would be rotated by $\pi/2$.

Since there are no impressed electric nor magnetic current densities it follows from (2.39) and (2.45) that the vector potentials obey the Helmholtz equation, that is,

$$\nabla^2 A_{z1} + k_1^2 A_{z1} = 0 \tag{5.7}$$
$$\nabla^2 A_{z0} + k_0^2 A_{z0} = 0 \tag{5.8}$$
$$\nabla^2 F_{z1} + k_1^2 F_{z1} = 0 \tag{5.9}$$
$$\nabla^2 F_{z1} + k_0^2 F_{z0} = 0 \tag{5.10}$$

where, as usual

$$k_1^2 = \omega^2 \varepsilon_1 \mu_0 \tag{5.11}$$
$$k_0^2 = \omega^2 \varepsilon_0 \mu_0 \tag{5.12}$$

Recalling (A.24) and substituting (5.3) in (5.7) we get after some manipulation

$$k_{\rho_1}^2 \rho^2 J_n''(k_{\rho_1}\rho) + k_{\rho_1}\rho J_n'(k_{\rho_1}\rho) + [\rho^2(k_1^2 - \beta^2) - n^2] J_n(k_{\rho_1}\rho) = 0 \tag{5.13}$$

where $J_n'(x)$ and $J_n''(x)$, respectively denote the first and second derivatives of the Bessel function of the first kind with respect to its argument.

Equation (5.13) is the Bessel equation and will only be satisfied for all values of ρ if

$$k_{\rho_1}^2 = k_1^2 - \beta^2 \tag{5.14}$$

Repeating the same procedure with (5.4) we get

$$k_{\rho_0}^2 = \beta^2 - k_0^2 \tag{5.15}$$

Had we taken (5.5) and (5.6) instead of (5.3) and (5.4) we would have also obtained (5.14) and (5.15).

Applying (2.61) to (2.64) we get the following expressions for the electromagnetic fields inside (subscript 1) and outside (subscript 0) the dielectric cylinder:

$$E_{\rho_1} = \frac{a_c}{\rho}\left[\frac{c_c n}{a_c \varepsilon_1} - \frac{\omega\beta}{k_1^2}\frac{k_{\rho_1}\rho J_n'(k_{\rho_1}\rho)}{J_n(k_{\rho_1}\rho)}\right] J_n(k_{\rho_1}\rho)\sin(n\varphi)\exp(\pm j\beta z) \tag{5.16}$$

$$E_{\rho_0} = \frac{b_c}{\rho}\left[\frac{d_c n}{b_c \varepsilon_0} - \frac{\omega\beta}{k_0^2}\frac{k_{\rho_1}\rho K_n'(k_{\rho_0}\rho)}{K_n(k_{\rho_0}\rho)}\right] K_n(k_{\rho_0}\rho)\sin(n\varphi)\exp(\pm j\beta z) \tag{5.17}$$

$$E_{\varphi_1} = \frac{a_c}{\rho}\left[\frac{c_c}{a_c \varepsilon_1}\frac{k_{\rho_1}\rho J_n'(k_{\rho_1}\rho)}{J_n(k_{\rho_1}\rho)} - \frac{\beta\omega n}{k_1^2}\right] J_n(k_{\rho_1}\rho)\cos(n\varphi)\exp(\pm j\beta z) \tag{5.18}$$

$$E_{\varphi_0} = \frac{b_c}{\rho}\left[\frac{d_c}{b_c \varepsilon_0}\frac{k_{\rho_0}\rho K_n'(k_{\rho_0}\rho)}{K_n(k_{\rho_0}\rho)} - \frac{\beta\omega n}{k_0^2}\right] K_n(k_{\rho_0}\rho)\cos(n\varphi)\exp(\pm j\beta z) \tag{5.19}$$

$$E_{z_1} = -j a_c \omega \frac{k_{\rho_1}^2}{k_1^2} J_n(k_{\rho_1}\rho)\sin(n\varphi)\exp(\pm j\beta z) \tag{5.20}$$

$$E_{z_0} = +j b_c \omega \frac{k_{\rho_0}^2}{k_0^2} K_n(k_{\rho_0}\rho)\sin(n\varphi)\exp(\pm j\beta z) \tag{5.21}$$

$$H_{\rho_1} = \frac{a_c}{\rho}\left[\frac{n}{\mu_0} - \frac{c_c}{a_c}\frac{\omega\beta}{k_1^2}\frac{k_{\rho_1}\rho J_n'(k_{\rho_1}\rho)}{J_n(k_{\rho_1}\rho)}\right] J_n(k_{\rho_1}\rho)\cos(n\varphi)\exp(\pm j\beta z) \quad (5.22)$$

$$H_{\rho_0} = \frac{b_c}{\rho}\left[\frac{n}{\mu_0} - \frac{d_c}{b_c}\frac{\omega\beta}{k_0^2}\frac{k_{\rho_0}\rho K_n'(k_{\rho_0}\rho)}{K_n(k_{\rho_0}\rho)}\right] K_n(k_{\rho_0}\rho)\cos(n\varphi)\exp(\pm j\beta z) \quad (5.23)$$

$$H_{\varphi_1} = \frac{a_c}{\rho}\left[\frac{c_c}{a_c}\frac{\omega\beta n}{k_1^2} - \frac{1}{\mu_0}\frac{k_{\rho_1}\rho J_n'(k_{\rho_1}\rho)}{J_n(k_{\rho_1}\rho)}\right] J_n(k_{\rho_1}\rho)\sin(n\varphi)\exp(\pm j\beta z) \quad (5.24)$$

$$H_{\varphi_0} = \frac{b_c}{\rho}\left[\frac{d_c}{b_c}\frac{\omega\beta n}{k_0^2} - \frac{1}{\mu_0}\frac{k_{\rho_0}\rho K_n'(k_{\rho_0}\rho)}{K_n(k_{\rho_0}\rho)}\right] K_n(k_{\rho_0}\rho)\sin(n\varphi)\exp(\pm j\beta z) \quad (5.25)$$

$$H_{z_1} = -jc_c\omega\frac{k_{\rho_1}^2}{k_1^2} J_n(k_{\rho_1}\rho)\cos(n\varphi)\exp(\pm j\beta z) \quad (5.26)$$

$$H_{z_0} = +jd_c\omega\frac{k_{\rho_0}^2}{k_1^2} K_n(k_{\rho_0}\rho)\cos(n\varphi)\exp(\pm j\beta z) \quad (5.27)$$

where $J_n'(x)$ denotes the first derivative of the Bessel function of first kind and $K_n'(x)$ the first derivative of the modified Bessel function of the second kind, each with respect to its argument.

Matching successively the E_z, H_z, E_φ, and H_φ field components at the cylinder sidewall $\rho = \rho_1$ we get

$$\frac{b_c}{a_c} = \frac{\varepsilon_0}{\varepsilon_1}\frac{\beta^2 - \omega^2\varepsilon_1\mu_0}{\beta^2 - \omega^2\varepsilon_0\mu_0}\frac{J_n(k_{\rho_1}\rho_1)}{K_n(k_{\rho_0}\rho_1)} = -\frac{\varepsilon_0}{\varepsilon_1}\frac{k_{\rho_1}^2}{k_{\rho_0}^2}\frac{J_n(k_{\rho_1}\rho_1)}{K_n(k_{\rho_0}\rho_1)} \quad (5.28)$$

$$\frac{d_c}{c_c} = -\frac{\varepsilon_0}{\varepsilon_1}\frac{k_{\rho_1}^2}{k_{\rho_0}^2}\frac{J_n(k_{\rho_1}\rho_1)}{K_n(k_{\rho_0}\rho_1)} \quad (5.29)$$

$$\frac{c_c}{a_c} = -\frac{(\varepsilon_0 - \varepsilon_1)\beta n\omega J_n(k_{\rho_1}\rho_1)K_n(k_{\rho_0}\rho_1)}{\rho_1\left[k_{\rho_1}k_{\rho_0}^2 J_n'(k_{\rho_1}\rho_1)K_n(k_{\rho_0}\rho_1) + k_{\rho_0}k_{\rho_1}^2 J_n(k_{\rho_1}\rho_1)K_n'(k_{\rho_0}\rho_1)\right]} \quad (5.30)$$

$$\frac{c_c}{a_c} = -\frac{\rho_1\left[\varepsilon_1 k_{\rho_1}k_{\rho_0}^2 J_n'(k_{\rho_1}\rho_1)K_n(k_{\rho_0}\rho_1) + \varepsilon_0 k_{\rho_0}k_{\rho_1}^2 J_n(k_{\rho_1}\rho_1)K_n'(k_{\rho_0}\rho_1)\right]}{(\varepsilon_0 - \varepsilon_1)\beta n\omega\mu_0 J_n(k_{\rho_1}\rho_1)K_n(k_{\rho_0}\rho_1)} \quad (5.31)$$

In the above we should note that

$$\frac{d_c}{b_c} = \frac{d_c}{c_c}\frac{c_c}{a_c}\frac{a_c}{b_c} \quad (5.32)$$

Substituting in (5.32) the value of d_c/c_c given in (5.29) and a_c/b_c from (5.28) we get

$$\frac{d_c}{b_c} = \frac{c_c}{a_c} \quad (5.33)$$

Introducing

$$F_n(x) = \frac{xJ_n'(x)}{J_n(x)} \tag{5.34}$$

$$M_n(x) = \frac{xK_n'(x)}{K_n(x)} \tag{5.35}$$

and equating the values of c_c/a_c given in (5.30) and (5.31) we get the characteristic equation for the hybrid modes in a dielectric cylinder:

$$\left[\frac{\varepsilon_1}{\varepsilon_0}\frac{F_n(k_{\rho_1}\rho_1)}{k_{\rho_1}^2} + \frac{M_n(k_{\rho_0}\rho_1)}{k_{\rho_0}^2}\right]\left[\frac{F_n(k_{\rho_1}\rho_1)}{k_{\rho_1}^2} + \frac{M_n(k_{\rho_0}\rho_1)}{k_{\rho_0}^2}\right] = \frac{(\varepsilon_0 - \varepsilon_1)^2\beta^2 n^2\omega^2\mu_0}{\varepsilon_0(k_{\rho_0}^2 k_{\rho_1}^2)^2} \tag{5.36}$$

Noting that

$$\begin{aligned}\left(\frac{1}{k_{\rho_1}^2} + \frac{1}{k_{\rho_0}^2}\right)^2 &= \left(\frac{\beta^2 - k_0^2 + k_1^2 - \beta^2}{k_{\rho_1}^2 k_{\rho_0}^2}\right)^2 \\ &= \left[\frac{\omega^2\mu_0(\varepsilon_1 - \varepsilon_0)}{k_{\rho_1}^2 k_{\rho_0}^2}\right]^2\end{aligned} \tag{5.37}$$

the second term in (5.36) can be modified and we finally get

$$\left[\frac{\varepsilon_1}{\varepsilon_0}\frac{F_n(k_{\rho_1}\rho_1)}{k_{\rho_1}^2} + \frac{M_n(k_{\rho_0}\rho_1)}{k_{\rho_0}^2}\right]\left[\frac{F_n(k_{\rho_1}\rho_1)}{k_{\rho_1}^2} + \frac{M_n(k_{\rho_0}\rho_1)}{k_{\rho_0}^2}\right] = \frac{\beta^2 n^2}{k_0^2}\left(\frac{1}{k_{\rho_1}^2} + \frac{1}{k_{\rho_0}^2}\right)^2 \tag{5.38}$$

For a given cylinder and each value of n the characteristic equation (5.38) has many solutions, each corresponding to a hybrid mode. Traditionally these modes are designated following the root order m as

- $\text{HE}_{n,l}$ for m odd;
- $\text{EH}_{n,l}$ for m even;

where

$$l = (m-1) \div 2 + 1 \tag{5.39}$$

and the symbol $\div$ stands for integer division. Because of its importance in dielectric rod antennas, the dominant $\text{HE}_{1,1}$ hybrid mode will be characterized here.

In Figure 5.3 we represent $\bar{\beta} = \beta/k_0$ as a function of the normalized cylinder radius, $\bar{\rho}_1 = k_0\rho_1$, for $\bar{\varepsilon}_1 = 2.53$, and 1.05. As expected $\bar{\beta}$ takes values between 1 and $\sqrt{\bar{\varepsilon}_1}$

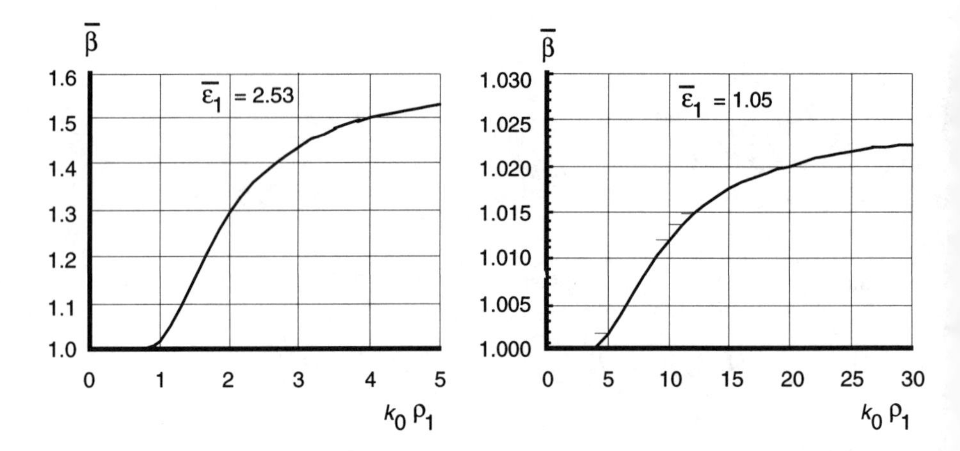

Figure 5.3: $\bar{\beta}$ as a function of $(k_0\rho_1)$ for the $HE_{1,1}$ mode of dielectric cylinders.

increasing with the cylinder radius. This justifies tapering down the cylinder to reduce reflections at the free end when the rod permittivity differs considerably from the free-space value.

We must note that the $HE_{1,1}$ has no cutoff, that is, it can propagate even when the cylinder radius approaches zero. As will become clear below, as the cylinder radius decreases the fields spread more outside the dielectric and become more difficult to excite.

The solution for (5.38) must be found numerically. However, there is a very simple procedure that provides an approximate solution adequate for many applications. In Section 3.4 we have shown that for the $TE^X_{1,1}$ mode of a dielectric waveguide with normalized rectangular cross section $\bar{a} \times \bar{b}$ $\bar{a} >> 1$ and $\bar{b} >> 1$ we have

$$\bar{\beta} = \sqrt{\bar{\varepsilon}_1 - \bar{k}_{x_1}^2 - \bar{k}_{y_1}^2} \tag{5.40}$$

where

$$\bar{k}_{x_1} \approx \frac{\pi}{\bar{a}}\left(1 - \frac{1}{1 + \frac{\bar{a}}{2}\sqrt{\bar{\varepsilon}_1 - 1}}\right) \tag{5.41}$$

$$\bar{k}_{y_1} \approx \frac{\pi}{\bar{b}}\left(1 - \frac{1}{1 + \frac{\bar{b}}{2}\sqrt{\bar{\varepsilon}_1 - 1}}\right) \tag{5.42}$$

If we now approximate the dielectric cylinder of radius $\bar{\rho}_1$ by a square waveguide of side $\bar{a}$ with the same cross section we have

$$\bar{a} = \sqrt{\pi}\,\bar{\rho}_1 \tag{5.43}$$

Using (5.43), substituting $\bar{a}$ for $\bar{\rho}_1$ in (5.41) and (5.42) and introducing the values of $\bar{k}_{x_1}$ and $\bar{k}_{y_1}$ in (5.40) we have

$$\bar{\beta} \approx \sqrt{\bar{\varepsilon}_1 - \frac{2\pi}{\bar{\rho}_1^2}\left[1 - \frac{1}{1 + \frac{\bar{\rho}_1}{2}\sqrt{\pi(\bar{\varepsilon}_1 - 1)}}\right]^2} \tag{5.44}$$

The values obtained using (5.44) have a relative error in $\bar{\beta}$ of the order of 10^{-2} or less for $\bar{\rho}_1 > 4$ and $\bar{\varepsilon}_1 \approx 2.5$, decreasing to less than $5 \cdot 10^{-4}$ for $\bar{\rho}_1 > 8$. The relative error in $\bar{\beta}$ is also less than 10^{-2} for $\bar{\rho}_1 > 5$ for and $\bar{\varepsilon}_1 \approx 1.05$ decreasing to less than 10^{-3} for $\bar{\rho}_1 > 10$. More important still, particularly for the low values of relative permittivity, the relative error in $\bar{\beta} - 1$ is less than 10^{-1} for $\bar{\rho}_1 > 10$ and less than 10^{-2} for $\bar{\rho}_1 > 20$.

5.3 HYBRID MODE FIELDS

For some applications it is more convenient to express the field values in rectangular coordinates. Choosing a system of rectangular coordinates centered on the cylinder axis, such that the axis OX coincides with $\varphi = 0$ we have

$$E_x = E_\rho \cos(\varphi) - E_\varphi \sin(\varphi) \tag{5.45}$$

$$E_y = E_\rho \sin(\varphi) + E_\varphi \cos(\varphi) \tag{5.46}$$

Substituting the values of E_ρ and E_φ given in (5.16) to (5.19) in (5.45) and (5.46), dropping an amplitude constant a_c and the factor $\exp[j(\omega t - \beta z)]$ and manipulating we get

$$E_{x_1} = \frac{\beta k_{\rho_1}}{2n\omega\varepsilon_1\mu_0}\{(\xi + n)J_{n+1}(k_{\rho_1}\rho)\sin[(n+1)\varphi] + (\xi - n)J_{n-1}(k_{\rho_1}\rho)\sin[(n-1)\varphi]\} \tag{5.47}$$

$$E_{x_0} = \frac{\beta k_{\rho_0}}{2n\omega\varepsilon_0\mu_0}\chi\,\{(\xi + n)K_{n+1}(k_{\rho_0}\rho)\sin[(n+1)\varphi)] - (\xi - n)K_{n-1}(k_{\rho_0}\rho)\sin[(n-1)\varphi]\} \tag{5.48}$$

$$E_{y_1} = \frac{\beta k_{\rho_1}}{2n\omega\varepsilon_1\mu_0}\{(\xi + n)J_{n+1}(k_{\rho_1}\rho)\cos[(n+1)\varphi] + (\xi - n)J_{n-1}(k_{\rho_1}\rho)\cos[(n-1)\varphi]\} \tag{5.49}$$

$$E_{y_0} = -\frac{\beta k_{\rho_0}}{2n\omega\varepsilon_0\mu_0}\chi\,\{(\xi + n)K_{n+1}(k_{\rho_0}\rho)\cos[(n+1)\varphi] + (\xi - n)K_{n-1}(k_{\rho_0}\rho)\cos[(n-1)\varphi]\} \tag{5.50}$$

Similarly for the H field we have

$$H_{x_1} = -\frac{k_{\rho_1}}{2n\mu_0}\{(\zeta+n)J_{n+1}(k_{\rho_1}\rho)\cos[(n+1)\varphi]+ (\zeta-n)J_{n-1}(k_{\rho_1}\rho)\cos[(n-1)\varphi]\} \quad (5.51)$$

$$H_{x_0} = -\frac{k_{\rho_0}\chi}{2n\mu_0}\left\{(n+\zeta\frac{\varepsilon_1}{\varepsilon_0})K_{n+1}(k_{\rho_0}\rho)\cos[(n+1)\varphi]- (\zeta\frac{\varepsilon_1}{\varepsilon_0}-n)K_{n-1}(k_{\rho_0}\rho)\cos[(n-1)\varphi]\right\} \quad (5.52)$$

$$H_{y_1} = -\frac{k_{\rho_1}}{2n\mu_0}\{(\zeta+n)J_{n+1}(k_{\rho_1}\rho)\sin[(n+1)\varphi)]- (\zeta-n)J_{n-1}(k_{\rho_1}\rho)\sin[(n-1)\varphi)]\} \quad (5.53)$$

$$H_{y_0} = -\frac{k_{\rho_0}\chi}{2n\mu_0}\left\{(\frac{\zeta\varepsilon_0}{\varepsilon_1}+n)K_{n+1}(k_{\rho_0}\rho)\sin[(n+1)\varphi]+ (\frac{\zeta\varepsilon_0}{\varepsilon_1}-n)K_{n-1}(k_{\rho_0}\rho)\sin[(n-1)\varphi]\right\} \quad (5.54)$$

where

$$\chi = -\frac{k_{\rho_1}^2 J_1(k_{\rho_1}\rho_1)}{k_{\rho_0}^2 K_1(k_{\rho_0}\rho_1)}\frac{\varepsilon_0}{\varepsilon_1} \quad (5.55)$$

$$\xi = \frac{k_1^2\zeta}{\beta^2} \quad (5.56)$$

$$\zeta = \frac{F_1(k_{\rho_1}\rho_1)+\frac{\varepsilon_0 k_{\rho_1}^2}{\varepsilon_1 k_{\rho_0}^2}M_1(k_{\rho_0}\rho_1)}{1+\frac{k_{\rho_1}^2}{k_{\rho_0}^2}} \quad (5.57)$$

The choice of a rectangular coordinate system may seem odd for a structure with cylindrical symmetry. It is, however, useful to show cross-polarization, both in the aperture and in the far field, and to use the Kirchhoff-Huygens radiation formulas as presented in Section 2.11.

For antenna work, particularly for the low permittivity dielectric cylinder, the most interesting modes are those with unity azimuthal dependence ($n = 1$). The following is restricted to these modes.

Inside the dielectric cylinder the $HE_{1,1}$ field is mostly linearly polarized, as shown in Figures 5.4 and 5.5. For the usual values of the permittivity cross polarized components have nonnegligible amplitudes outside the dielectric while for low permittivities even if cross-polarization increases outside the dielectric the maximum values are still very low. The same type of result is well known for low-contrast optical fibers.

In Figure 5.6 we represent $H_x(\rho/\rho_1, \varphi = \pi/2)$ normalized to the value on axis. When the frequency, or the cylinder radius measured in wavelengths, increases the fields decay

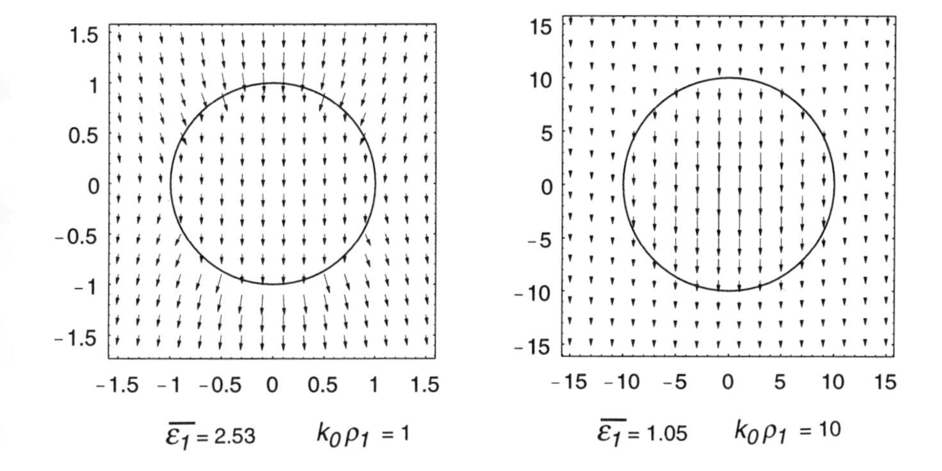

Figure 5.4: Field lines of the $HE_{1,1}$ mode in dielectric cylinders.

faster outside the cylinder. Here we should notice that in typical dielectric rod antennas ($\bar{\varepsilon}_1 \approx 2.5$ and $k_0\rho_1 < 0.5$) the field extends way out of the dielectric and the field value at the dielectric boundary is relatively high (> -3 dB) while for usual dielectric foam cylinder antennas ($\bar{\varepsilon}_1 \approx 1.05$ and $k_0\rho_1 > 10$) the field tend to be highly concentrated inside the dielectric and field intensities at the boundary are below -7 to -10 dB.

The distribution of the power in the region 0 to ρ, normalized to the total mode power, for various cylinder sizes and permittivities is plotted in Figure 5.7. This figure shows again the difference between typical dielectric rod antennas where a large fraction of the power (often more than 50%) is carried outside the dielectric and dielectric foam cylinder antennas where, despite the low relative permittivity values, most of the power (typically more than 90%) is carried inside the cylinder.

5.4 RADIATION PATTERN OF THE $HE_{1,1}$ MODE

The radiation pattern of a dielectric cylinder antenna may be calculated using the Kirchhoff-Huygens formulas and integrating over a suitably chosen aperture where the fields are specified accurately enough. As in the dielectric waveguide antenna, we choose an aperture comprising the dielectric free end and the sidewall. Assuming the fields to be zero elsewhere this procedure would produce exact results if we were able to specify exactly the fields on the aperture. Since this is a rather difficult task, at least in the general case, we will approximate the fields on the aperture by the unperturbed incident $HE_{1,1}$ mode fields and if required add to the resulting radiation pattern the pattern of the launcher as in Section 3.8.

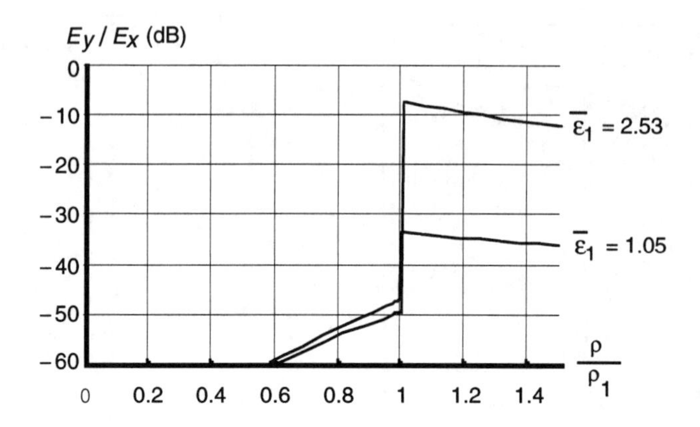

Figure 5.5: Worst-case ($\varphi = \pi/4$) aperture cross-polarization level for a low-permittivity dielectric cylinder ($\bar{\varepsilon}_1 = 1.05$, $k_0\rho_1 = 10$).

After considerable manipulation we arrive at the following results, for the far-field components of the radiation pattern at the distance R of a dielectric cylinder with length ℓ operated on the $\mathrm{HE}_{1,1}$ mode:

$$\begin{aligned}
E_\theta^{\text{side}}(\theta,\phi) &= \frac{jk\exp(-jkR)}{2R}\rho_1\sin(\phi) \\
&\int_0^\ell \left\{\left[jZH_{\varphi_1}(\rho_1,z)\sin(\theta) + \frac{ZH_{z_1}(\rho_1,z)}{k\rho_1\sin(\theta)}\cos(\theta)\right]J_1(k\rho_1\sin\theta)\right. \\
&\left. -E_{z_1}(\rho_1,z)J_1'(k\rho_1\sin\theta)\right\}\exp(jkz\cos\theta)dz \qquad (5.58)
\end{aligned}$$

$$\begin{aligned}
E_\phi^{\text{side}}(\theta,\phi) &= \frac{jk\exp(-jkR)}{2R}\rho_1\cos(\phi) \\
&\int_0^\ell \left\{\left[jE_{\varphi_1}(\rho_1,z)\sin(\theta) - \frac{E_{z_1}(\rho_1,z)}{k\rho_1\sin(\theta)}\cos(\theta)\right]J_1(k\rho_1\sin\theta)+\right. \\
&\left. ZH_{z_1}(\rho_1,z)J_1'(k\rho_1\sin\theta)\right\}\exp(jkz\cos\theta)dz \qquad (5.59)
\end{aligned}$$

$$\begin{aligned}
E_\theta^{\text{end}}(\theta,\phi) &= \frac{jk\exp[-jk(R-\ell\cos(\theta)]}{2R}\sin(\phi) \\
&\int_0^{\rho_1}\left\{[E_{\varphi_1}(\rho,\ell) - ZH_{\rho_1}(\rho,\ell)\cos(\theta)]\frac{J_1(k\rho\sin\theta)}{k\rho\sin\theta}+\right. \\
&\left.[E_{\rho_1}(\rho,\ell) + ZH_{\varphi_1}(\rho,\ell)\cos(\theta)]\,J_1'(k\rho\sin\theta)\right\}rdr \qquad (5.60)
\end{aligned}$$

$$\begin{aligned}
E_\phi^{\text{end}}(\theta,\phi) &= \frac{jk\exp[-jk(R-\ell\cos(\theta)]}{2R}\sin(\phi) \\
&\int_0^{\rho_1}\{[E_{\varphi_1}(\rho,\ell)\cos(\theta) - ZH_{\rho_1}(\rho,\ell)]\,J_1'(k\rho\sin\theta)+
\end{aligned}$$

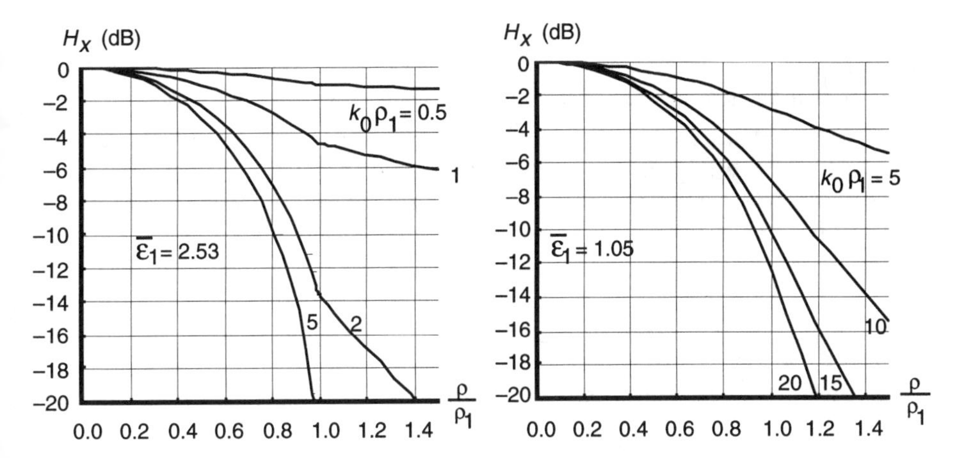

Figure 5.6: Computed magnetic field of the $HE_{1,1}$ mode of dielectric cylinders of varying size as a function of ρ/ρ_1 for $\bar{\varepsilon}_1 = 2.53$ and 1.05. All patterns are normalized to boresight.

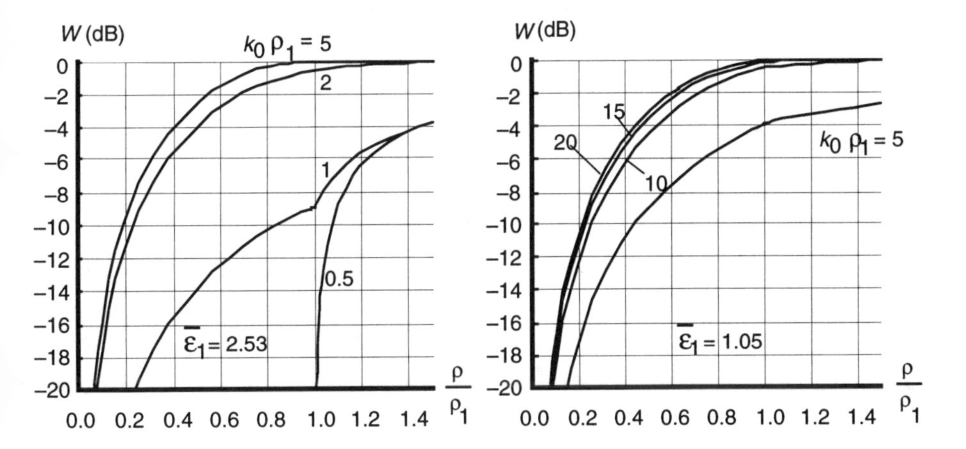

Figure 5.7: Computed power in the region $\rho = 0$ to ρ as a fraction to the total mode power for dielectric cylinders with $\bar{\varepsilon}_1 = 2.53$ and 1.05 varying diameters.

$$[E_{\rho_1}(\rho,\ell)\cos(\theta) + ZH_{\varphi_1}(\rho,\ell)]\,\frac{J_1(k\rho\sin\theta)}{k\rho\sin\theta}\Bigg\}\,rdr \tag{5.61}$$

where E_{φ_1}, E_{z_1}, H_{φ_1}, and H_{z_1} are given in (5.18), (5.20), (5.24), and (5.26), respectively, eliminating the φ dependence that has already been dealt with in the integration.

Figure 5.8 shows the computed radiation pattern for $\bar{\varepsilon}_1 = 2.63$, $\bar{\rho}_1 = 1.225$, and $\bar{\ell} = 23.37$. For typical dielectric rod antennas, such as this one, the radiation pattern is highly dependent on the rod length, exhibits good circular symmetry but, just like the dielectric waveguide antenna, has an undesirable high-sidelobe level.

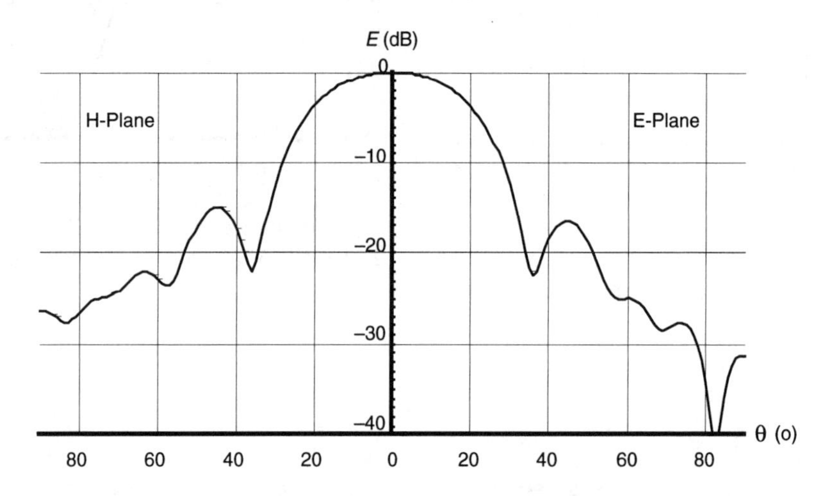

Figure 5.8: Computed far-field radiation pattern of a dielectric rod antenna with $\bar{\varepsilon}_1 = 2.63$, $\bar{\rho}_1 = 1.225$, and $\bar{\ell} = 23.37$ operated in the $HE_{1,1}$ mode. Launcher effects are not included.

The situation is very different with the usual dielectric foam cylinder antennas, such as the one shown in Figure 5.9, for $\bar{\varepsilon}_1 = 1.045$, $\bar{\rho}_1 = 11.5$, and $\bar{\ell} = 73.7$. Now the free end to side wall area ratio increases and, at the same time, the fields at the cylinder wall decrease. Acting together these two factors reduce the importance of the sidewall in the radiation pattern which becomes much more dependent on the free end and therefore much less dependent on the cylinder length.

5.5 EXCITATION OF THE $HE_{1,1}$ MODE

Consider a semi-infinite dielectric cylinder with ε_1 and μ_0 immersed in an otherwise free half-space with ε_0 and μ_0 and a coordinate system as shown in Figure 5.2.

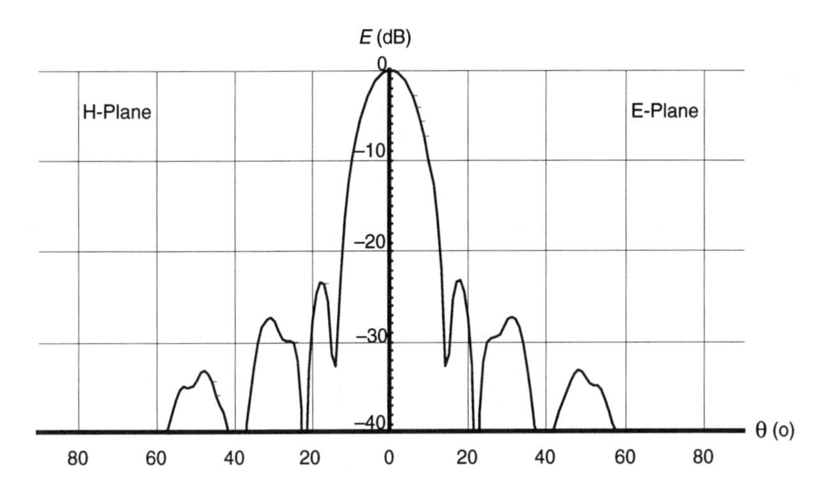

Figure 5.9: Computed far-field radiation pattern of a dielectric rod antenna with $\bar{\varepsilon}_1 = 1.045$, $\bar{\rho}_1 = 11.5$, and $\bar{\ell} = 73.7$ operated in the $HE_{1,1}$ mode. Launcher effects are not included.

Assume that an electromagnetic field structure $\boldsymbol{E}_i, \boldsymbol{H}_i$ impinges on the left-hand side ($z < 0$) of plane XOY and that this is the only field structure that can exist on that half space characterized by ε_0 and μ_0.

This problem is similar to the one described in Section 3.6, the only difference being the shape of the dielectric structure. Proceeding as before we calculate the excitation efficiency $\eta_{m,n}$ of any given mode as the ratio between the power in that mode and the incident power using (3.131).

Omitting a common factor $\exp[j(\omega t - \beta_m z)]$, the $TE_{1,1}$ fields of a metallic circular waveguide with radius ρ_m, filled with the same material as the dielectric cylinder, are given by

$$E_{i_\rho} = \frac{J_1(k_m\rho}{\rho}) \sin(\varphi) \tag{5.62}$$

$$E_{i_\varphi} = J_1'(k_m\rho) \cos(\varphi) \tag{5.63}$$

$$H_{i_\rho} = -\frac{\beta_m\sqrt{\bar{\varepsilon}_1}}{Z_0} J_1'(k_m\rho) \cos(\varphi) \tag{5.64}$$

$$H_{i_\varphi} = \frac{\beta_m\sqrt{\bar{\varepsilon}_1}}{Z_0} \frac{J_1(k_m\rho}{\rho}) \sin(\varphi) \tag{5.65}$$

where $k_m\rho_m$ is the smallest root of the equation

$$J_1'(k_m\rho_m) = 0 \tag{5.66}$$

In Figure 5.10 we plot the excitation efficiency of the $HE_{1,1}$ mode fields in dielectric cylinders, with relative permittivity $\bar{\varepsilon}_1 = 2.53$ and 1.05 and normalized radius $k_0\rho_1 = 1$, 2, and 5 for the first case and 5, 10, and 20 for the second case, using as launcher a circular metallic waveguide, filled with the same dielectric, with internal radius $k_0\rho_m$ and operated in the $TE_{1,1}$ mode. As shown the excitation efficiency can be quite high (≈ 0.8) over a reasonably large bandwidth. Maximum efficiency is always achieved with a metallic waveguide radius slightly larger than the dielectric cylinder radius, although this difference tends to vanish as the latter increases.

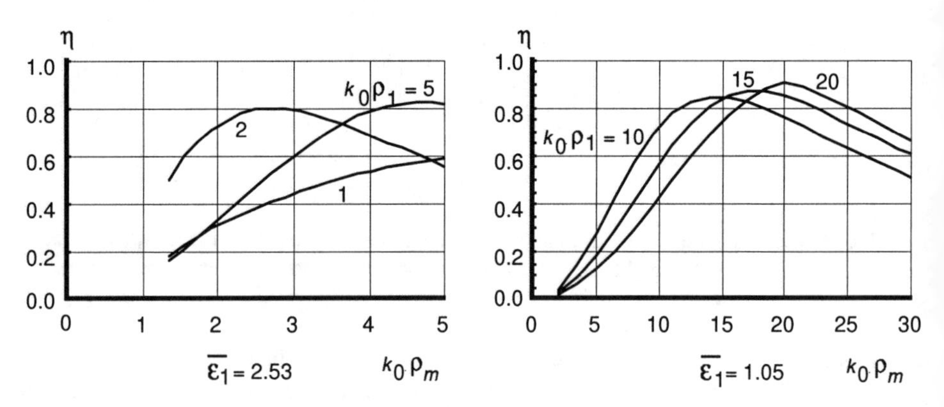

Figure 5.10: Excitation efficiency of the $HE_{1,1}$ mode of dielectric cylinders with radius $k_0\rho_1$ using as the launcher a circular metallic waveguide filled with the same dielectric, with internal radius $k_0\rho_m$ operated in the $TE_{1,1}$ mode.

The use of the following other launchers has been dealt with elsewhere [18] and will not be repeated here:

- $TE_{1,0}$ mode of a square metallic waveguide;
- $HE_{1,1}$ balanced hybrid mode of a corrugated metallic circular waveguide.

From the available results three important practical conclusions can be drawn:

1. For a primitive launcher, such as the square waveguide, the excitation efficiency can reach 80%; this value increases to 85% to 90% for a circular waveguide and closely approaches 100% for a corrugated circular waveguide.
2. Due to the spreading of the fields outside the dielectric rod the optimum efficiency launcher always has a larger cross section than the dielectric rod.
3. The lower the frequency and the dielectric permittivity the larger the ratio between the launcher and the rod cross section for maximum efficiency (and usually the lower the maximum efficiency).

It can be shown that the combination is broadband and bandwidths of an octave may be achieved with efficiencies close to the maximum. As a typical example results for a dielectric cylinder with 12.7 cm diameter and $\bar{\varepsilon}_1 = 1.05$ excited by a 15.2 cm diameter circular waveguide are given in Figure 5.11.

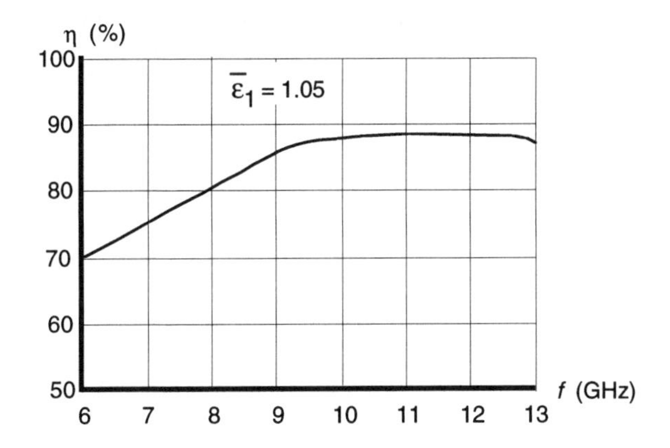

Figure 5.11: Excitation efficiency of the $HE_{1,1}$ mode of a dielectric cylinder with $\bar{\varepsilon}_1 = 1.05$ and 12.7-cm diameter using as the launcher a metallic circular waveguide with a diameter of 15.2 cm operated in the $TE_{1,1}$ mode.

5.6 EFFECT OF LAUNCHER

When a dielectric cylinder, operated in the $HE_{1,1}$ dominant hybrid mode, and its launcher are used as an antenna, the observed radiation pattern is not only due to the dielectric cylinder but also to the launcher.

For relative permittivities of about 2.5, the method described in Section 3.8 may be used. The launcher is assumed to radiate unaffected by the presence of the dielectric cylinder, except for a constant amplitude factor due to the energy converted into the hybrid mode and an offset of its phase centre launcher, in the negative direction of the Z axis.

A rigorous account of the launcher effect may be found in [19, 20] for planar structures and in [21] for cylindrical structures. There it is shown that the cylinder modes contribute to the radiation inside a "trapping region," defined as $\theta <= \theta_T$, and the continuous spectrum wave due to the launcher exists everywhere but predominates outside "trapping region". From [21] it can be shown that the trapping angle θ_T is given by

$$\theta_T = \arccos\left\{\frac{1}{\cosh\left[0.5\log_e\left(\frac{1+\frac{k\rho_0}{\beta}}{1-\frac{k\rho_0}{\beta}}\right)\right]}\right\} \tag{5.67}$$

For very low relative permittivities (of the order of 1.2 or less) and cylinder radius larger that about one wavelength, the cylinder radiation pattern is much narrower than that of the launcher and is virtually sidelobe free. In this case the rigorous analysis supports the following simplified procedure. Take the far-field radiation pattern of the launcher in the absence of the dielectric cylinder (see Figure 5.12) and identify the "trapping region."

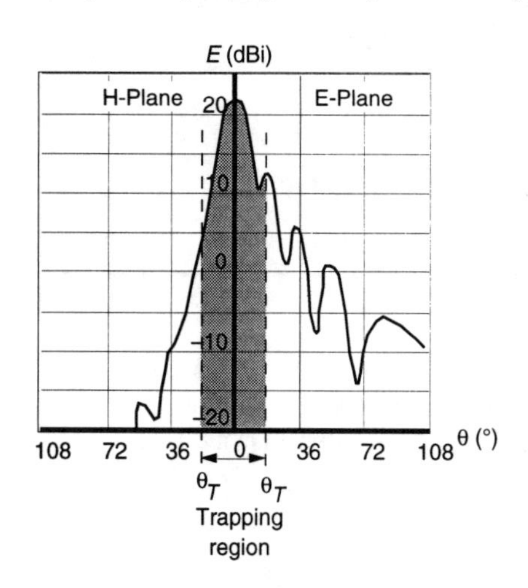

Figure 5.12: Computed far-field radiation pattern at 9 GHz of a pyramidal horn used later as a launcher.

Inside the trapping region the radiation will be due exclusively to the cylinder modes. Outside, the radiation pattern will be taken as the unaffected launcher pattern (see Figure 5.13). There remains an ill-defined region, for values of θ close to θ_T where the radiation pattern cannot be predicted by such a simple theory.

In the simplified case the "trapping region" may be easily explained, in terms of geometrical optics. If, in (5.67), we take the limit of θ_T for large values of $k_0\rho_1$ we get

$$\begin{aligned}\theta_T &= \frac{\pi}{2} - \arcsin\left(\frac{1}{\sqrt{\varepsilon_1}}\right) \\ &= \arccos\left(\frac{1}{\sqrt{\varepsilon_1}}\right)\end{aligned} \tag{5.68}$$

where θ_T may be seen as the complementary of the critical angle. Experimental results (Figures 5.14 to 5.16) give reasonable support to this method.

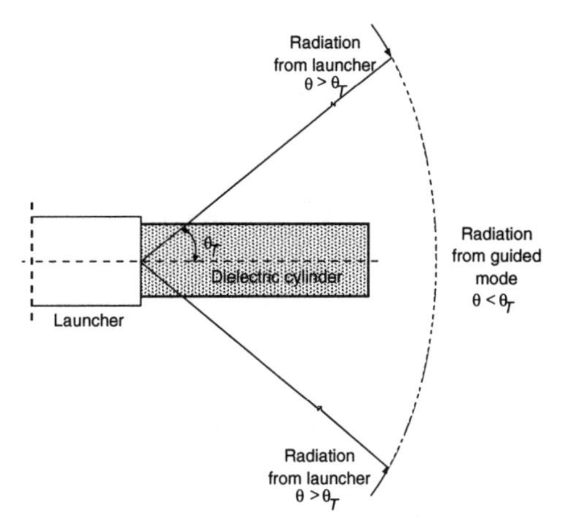

Figure 5.13: Schematic diagram illustrating the simplified theory to predict the radiation pattern of a low-permittivity dielectric cylinder antenna.

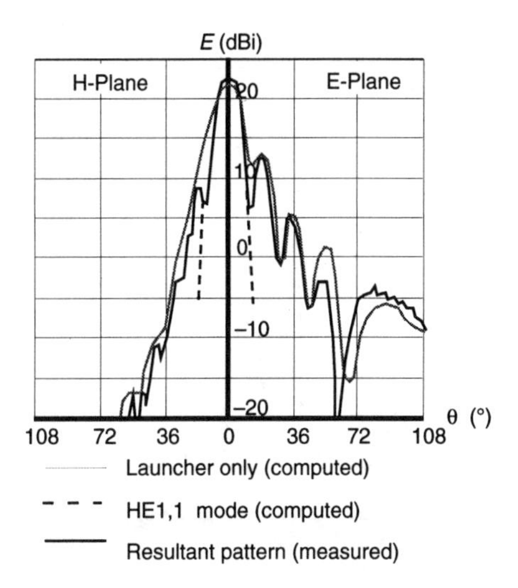

Figure 5.14: Far-field radiation pattern of a low-permittivity dielectric cylinder antenna, with $\bar{\varepsilon_1} = 1.045$ and $k_0\rho_1 = 11.5$ and normalized length 73.7 excited by a pyramidal horn.

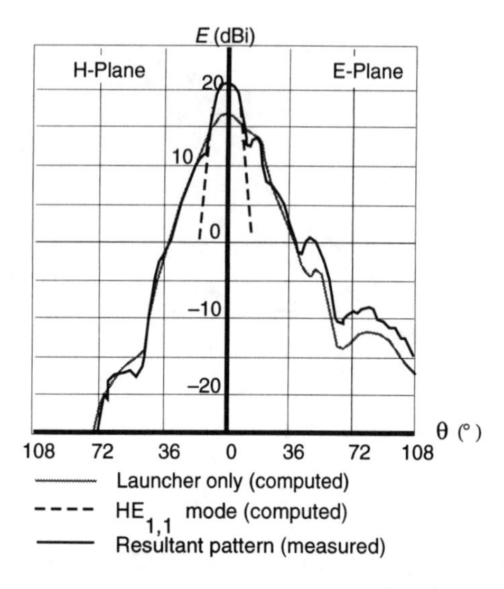

Figure 5.15: Far-field radiation pattern of a low-permittivity dielectric cylinder antenna, with $\bar{\varepsilon}_1 = 1.045$ and $k_0\rho_1 = 11.5$ and normalized length 73.7 excited by a conical horn with normalized mouth radius 14.5 and semiflare angle 30 degrees.

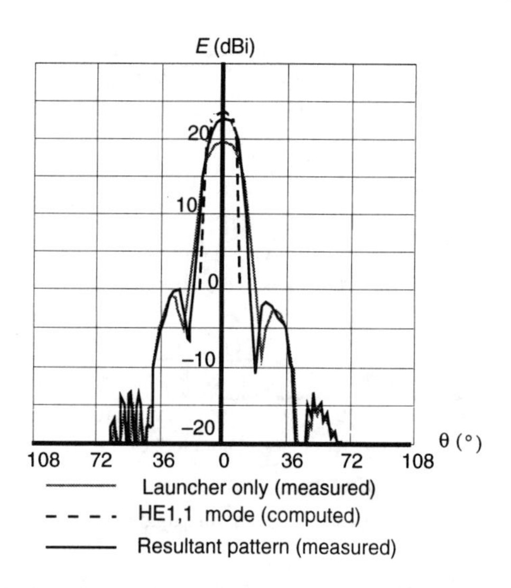

Figure 5.16: Far-field radiation pattern of a low-permittivity dielectric cylinder antenna, with $\bar{\varepsilon}_1 = 1.045$ and $k_0\rho_1 = 11.5$ and normalized length 73.7 excited by a corrugated waveguide with normalized mouth internal radius 11.9.

Where calculation of the energy content in the trapping region is not practical computed excitation efficiencies may be used instead.

In the examples that follow, with one exception, the launchers do not possess a plane equiphase surface and therefore efficiencies smaller than predicted for plane equiphase surfaces should be expected. An estimation of the loss in excitation efficiency due to the phase errors may be made from the graphs provided in Chapter 6.

So far we have only considered the presence of the $HE_{1,1}$ mode in the cylinder. In cases where its normalized radius $k_0\rho_1$ is such that [22]

$$\bar{\varepsilon}_1 > 1 + \left(\frac{2.4048}{k_0\rho_1}\right)^2 \tag{5.69}$$

the first higher-order modes ($TE_{0,1}$, $TM_{0,1}$, and $HE_{2,1}$) can propagate. These modes however will not be excited by launchers for which the field dependence in φ is of the form $\cos(\varphi)$ or $\sin(\varphi)$, such as it is in the dominant mode of circular waveguides (smooth or corrugated). The next higher-order mode that can be excited is the $EH_{1,1}$, whose cutoff condition may be written as [22]

$$\bar{\varepsilon}_1 > 1 + \left(\frac{3.8317}{k_0\rho_1}\right)^2 \tag{5.70}$$

5.7 SHIELDED DIELECTRIC CYLINDER

The variation of the the $HE_{1,1}$ mode transverse fields in a low-permittivity dielectric cylinder with the cylinder radius closely resemble a Gaussian curve, as will be shown below in detail. The radiation pattern of its free end should therefore be expected to exhibit very low sidelobes in addition to good circular symmetry.

If we enclose the dielectric cylinder and its launcher in a suitably sized metallic shield, very much in the same way as described for the previous shielded antennas (see Sections 3.9 and 4.7), we can rather effectively block the influence of the sidewalls and the launcher on the radiation pattern, which then is simply due to the cylinder free end. Besides its electromagnetic function the shield may double as an environmental protection of the dielectric foam cylinder.

Assuming that radiation is only due to the free end aperture and taking the aperture fields as the unperturbed incident fields, which is a reasonable assumption because the modal and the free-space impedances are very close, the following radiation patterns are obtained (Figure 5.17). For ease of comparison all patterns are normalized to boresight, and the normalizing factor is indicated in Table 5.1.

Table 5.1: Boresight directivity, in dBi, for the set of dielectric cylinders whose E-plane radiation patterns are represented in Figure 5.17.

Dielectric constant $\bar{\varepsilon}_1$	Normalized radius $k_0\rho_1$	Directivity (dBi)
1.05	8	25.17
1.05	10	25.03
1.05	12	25.13
1.05	14	25.58
1.05	16	26.07

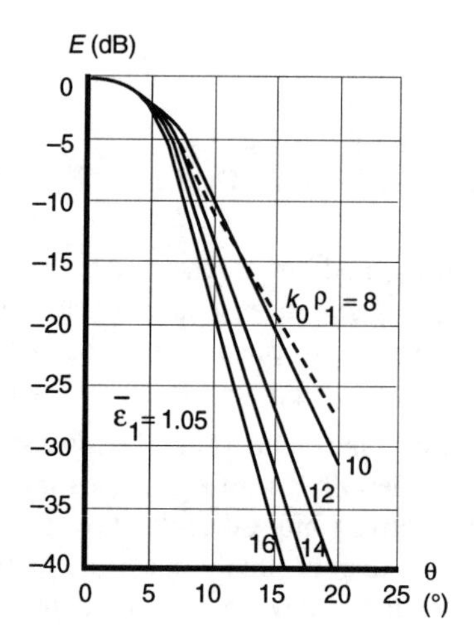

Figure 5.17: Computed E-plane far-field radiation patterns of dielectric cylinders with $\bar{\varepsilon}_1 = 1.05$ operated in the $HE_{1,1}$ mode. The patterns are normalized to yield the same boresight directivity.

When the frequency, or the cylinder radius for a given frequency, increases, the fields become more closely bound to the cylinder partly compensating the increasing normalized aperture size. Thus, for appropriate cylinder radius the radiation patterns exhibit smaller dependence on frequency than would otherwise be expected.

Although the low-permittivity shielded cylinder has many desirable properties that should make it a very useful antenna, in practice it presents a major drawback: it requires a large-sized launcher. As we showed in Section 5.5, to achieve an efficient excitation of the $HE_{1,1}$ mode the launcher cross-sectional dimensions should be slightly larger than the dielectric cylinder, which itself is already rather large ($k_0\rho_1 > 10$). In addition, a large launcher if not properly designed may either exhibit its own unwanted higher-order modes or may excite higher-order modes in the dielectric cylinder.

5.8 RADIATION PATTERN ESTIMATION

To estimate the radiation pattern of a shielded dielectric cylinder antenna operated in the $HE_{1,1}$ mode, one may use a few very simple methods. Possibly the most accurate one assumes that the aperture fields may be approximated by a Gaussian beam. Following Deschamps [23], let us use the paraxial method of analysis, where a sphere of radius R is substituted by a paraboloid of the same curvature (at the axis), and assume that all the sources are located in the region of $z < 0$. Referring to Figure 5.2 and taking the field f at $z = 0$ where we omit the time variation $\exp(j\omega t)$ as

$$
\begin{aligned}
f(\rho, 0) &= \exp\left(-\frac{\rho^2}{2\sigma_0^2}\right) \\
&= \exp\left(-\frac{k_0 x^2}{2a}\right)
\end{aligned}
\tag{5.71}
$$

we can write

$$
f(\rho, z) = \frac{1}{1 - j\frac{z}{a}} \exp\left(-\frac{k_0\rho^2}{2A}\right) \exp\left[-jk_0\left(z - \frac{\rho^2}{2R}\right)\right] \tag{5.72}
$$

where

$$
A = a + \frac{z^2}{a} \tag{5.73}
$$

$$
R = z + \frac{a^2}{z} \tag{5.74}
$$

For convenience let us rewrite (5.71) as

$$f(\rho, 0) = \exp\left[-\frac{k_0\rho_1^2}{2a} \cdot \left(\frac{\rho}{\rho_1}\right)^2\right] \tag{5.75}$$

where ρ_1 is a constant.

Manipulating (5.75) we have

$$\sqrt{-\log[f(\rho, 0)]} = b\frac{\rho}{\rho_1} \tag{5.76}$$

where

$$b = \frac{k_0\rho_1}{\sqrt{2ak_0}} \tag{5.77}$$

Taking $f(\rho, 0)$ as $H(\rho/\rho_1, 0)$ and plotting $y = \sqrt{-\log_e H(\rho/\rho_1, 0)}$ as a function of ρ/ρ_1 (as the example in Figure 5.18 shows) we conclude that a straight line is a reasonably good approximation of $y(\rho/\rho_1)$ meaning that $H(\rho/\rho_1, 0)$ may indeed be approximated by a Gaussian curve.

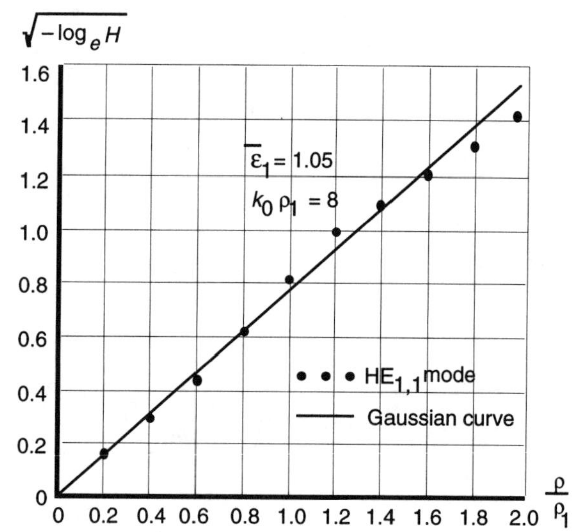

Figure 5.18: Test for Gaussian character of the $\text{HE}_{1,1}$ mode of a dielectric cylinder with $\bar{\varepsilon}_1 = 1.05$ and $k_0\rho_1 = 8$.

Fitting a straight line to the function $y(\rho/\rho_1)$ and making use of (5.77) we obtain a value for b for each value of $\bar{\varepsilon}_1$ and $k_0\rho_1$. For convenience we plot in Figure 5.19 $\sqrt{2ak_0}$ as a function of $k_0\rho_1$, for relative permittivity values of 1.05, 1.10, and 2.53.

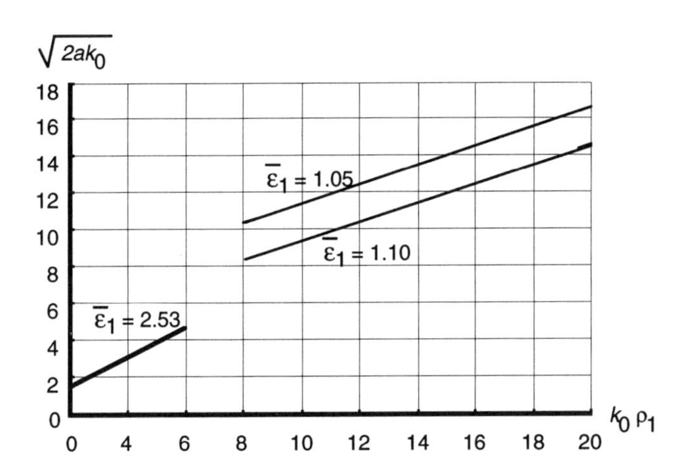

Figure 5.19: Value of $(2ak_0)^{1/2}$ as a function of $k_0\rho_1$ for $\bar{\varepsilon}_1 = 1.05$ and 1.10.

Further simplifications of (5.72) are often justified. For instance, if we calculate the fields at a distance z_1 from the aperture as a function of θ, it is possible to write (for small values of θ)

$$z = z_1 \cos\theta \tag{5.78}$$

$$\sin\theta \approx \theta \tag{5.79}$$

$$\left[\frac{f(\theta, z_1)}{f(0, z_1)}\right]_{dB} = -8.686\frac{(k_0 z_1)^2\theta^2}{2Ak_0} \tag{5.80}$$

$$\arg\left[\frac{f(\theta, z_1)}{f(0, z_1)}\right] = -\frac{x^2}{2Az_1} - \frac{k_0 x^2}{2z_1} + \frac{k_0 x^2}{2R} \tag{5.81}$$

where

$$A = a + \frac{z_1^2}{a} \tag{5.82}$$

$$R = z_1 + \frac{a^2}{z_1} \tag{5.83}$$

Boresight directivity referred to an isotropic source can also be easily calculated:

$$di_{dBi} = 20\log_{10}\left(\sqrt{\frac{4k_0 z_1^2}{A}}\right) \tag{5.84}$$

In the far field, when $z_1^2/a \gg a$ we have simply

$$
\begin{aligned}
di_{dBi} &= 10\log_{10}(4k_0a) && (5.85)\\
\left[\frac{f(\theta)}{f(0)}\right]_{dB} &= -4.343k_0a\theta^2 && (5.86)\\
\arg\left[\frac{f(\theta)}{f(0)}\right] &= 0 && (5.87)
\end{aligned}
$$

where θ as usual is expressed in radians.

Rewriting the expression for the directivity (5.85) as making use of (5.71) we have

$$
\begin{aligned}
di_{dBi} &= 10\log_{10}(4k_0a) && (5.88)\\
&= 10\log_{10}(4k_0^2\sigma_0^2) && (5.89)\\
&= 20\log_{10}(2k_0\sigma_0) && (5.90)
\end{aligned}
$$

Given the directivity di_u of a uniformly illuminated circular aperture of radius ρ_1

$$di_{u_{dBi}} = 20\log_{10}(k_0\rho_1) \quad (5.91)$$

we can derive the following relation between ρ_1 and σ_0:

$$\rho_1 = 2\sigma_0 \quad (5.92)$$

Comparing the values obtained with (5.86) and with a Kirchhoff-Huygens vector integration of the aperture field we find that amplitude errors are usually limited to about 1 dB, for a dynamic range of about 20 dB. Boresight directivity calculated by (5.85) and by integration of the aperture fields compare even better for the range of values of $k_0\rho_1$ displayed.

5.9 CONCLUSIONS

A finite-length dielectric cylinder operated in the dominating $HE_{1,1}$ mode may be used as an antenna. Many metallic antennas may be used as launchers but the most usual is possibly the circular waveguide.

For the usual values of permittivity ($\bar{\varepsilon}_1 \approx 2.5$) a small cylinder radius ($k_0\rho_1 \approx 1$) and a moderate cylinder length ($k_0\ell$ in the range 20 to 40) provide medium gain (13 to 16 dBi) and a reasonably circular symmetric radiation pattern. Sidelobe level, however, may prove to be too high for many applications.

Much lower values of permittivity ($\bar{\varepsilon}_1 \approx 1.1$) together with much higher cylinder radius ($k_0\rho_1 \approx 10$) lead to higher gains ($\approx$ 25 dBi), improved circular symmetry, lower cross-polarization, and lower sidelobe levels. The penalty to pay is a large launcher that, if not properly designed, may either present unwanted high-order modes or excite modes other than the $HE_{1,1}$.

The effect of launcher and the sidewall detracts on the performance that could be obtained from the low-permittivity dielectric cylinder antenna, if its radiation pattern was only due to the dielectric cylinder free end. The shielded low-permittivity dielectric cylinder antenna achieves just that objective and, as a result, exhibits low or very low sidelobe levels, coupled with good circular symmetry and low cross-polarization. In practice the large launcher size often remains an obstacle.

When the dielectric is shaped into a cone, the radiating aperture is no longer of the same size as the exciting aperture and smaller launchers become practical. The dielectric cone structure exhibits very useful radiation characteristics and will be the subject of the next chapter.

REFERENCES

[1] Chattterjee, R., (1985), *Dielectric and Dielectric-Loaded Antennas*, Research Study Press, Letchworth, Hertfordshire, England.

[2] Hondros, D., and Debye, P., (1910), Electromagnetische Wellen an Dielektrischen Drahten, *Annales der Physik*, Vol. 32, No. 8, pp. 465–476.

[3] James, J. R., (1968), *Studies of Cylindrical Dielectric-Rod Antennas*, The Royal Military College of Science, Department of Electrical and Instrument Technology, Technical Note RT41.

[4] James, J. R., (1967), Theoretical Investigations of Cylindrical Dielectric Rod Antennas, *Proceedings of the IEE*, Vol. 114, pp. 309–319.

[5] Horton, C. W., Karal, F. C., and McKinney, C., (1949), On the Radiation Patterns of Dielectric Rods of Circular Cross Section-TM01 mode, *Journal of Applied Physics*, Vol. 21, pp. 1279–1283.

[6] Bouix, M., (1952), *Contribution à l'ètude des antennes diélectriques*, Thése de Doctorat, Université de Paris.

[7] Fradin, A. Z., (1961) *Microwave Antennas*, Pergamon Press.

[8] Brown, J., and Spector, J. O., (1957), The Radiation Properties of End-fire Aerials, *Proceedings of the IEE* Vol. 104B, pp. 27–34.

[9] Newmann, E. G., (1970), Radiation Mechanism of Dielectric Rod and Yagi Aerials, *Electronics Letters*, Vol. 6, No. 16, pp. 528–530.

[10] Andersen, J. B., (1971), *Metallic and Dielectric Antennas*, Polyteknisk Forlag, Danmarks Tekniske Hojskole, Lyngby.

[11] Yaghjian, A. D., and Kornhauser, E. T., (1972), A Modal Analysis of the Dielectric Rod Antenna Excited by the $HE_{1,1}$ Mode, *Transactions of the IEEE* , Vol. AP-20, No. 2, pp. 122–128.

[12] Collin, R. E., (1960), *Field Theory of Guided Waves*, McGraw-Hill, New York.

[13] Angulo, C. M., and Chang, W., (1958), The Excitation of a Dielectric Rod by a Cylindrical Waveguide, *IRE Transactions on Microwave Theory and Techniques*, Vol. MTT-6, pp. 389–393.

[14] Olver, A. D., Clarricoats, P. J. B., Kishk, A. A., and Shafai, L., (1994), *Microwave Horns and Feeds*, IEEE Press, New York.

[15] Lewis, B. L., (1966), Large Diameter Dielectric Rod End-fire Antennas, *IEEE Transactions on Antennas and Propagation*, Vol. AP-14, No. 2, pp. 239–240.

[16] Clarricoats, P. J. B., (1970), Similarities in the Electromagnetic Behavior of Optical Waveguides and Corrugated Fields, *Electronics Letters*, Vol. 6, Nr. 6, pp.178–180.

[17] Clarricoats, P. J. B., (1960), Propagation Along Unbounded and Bounded Dielectric Rods, *The Institute of Electrical Engineers. Monograph No. 410E*, London.

[18] Clarricoats, P. J. B. and Salema, C. (1970), Design of Dielectric Cone Feeds for Microwave Antennas, *Proceedings of the 1971 European Microwave Conference*, Vol. 1, Paper B5/4.

[19] Tamir, T., and Oliner, A., (1963), Guided Complex Waves (Part I and Part II) *Proceedings of the IEE*, Vol. 110, No. 2, pp. 310–334.

[20] Shevchenko, V., (1971), *Continuous Transitions in Open Waveguides*, Golem Press.

[21] Fernandes, C., and Barbosa A., (1990), Complex Wave Radiation from a Sheath Helix Excited by the Circular Waveguide TE_{11} Mode, *Journal of Electromagnetic Waves and Applications*, Vol. 4, No. 6, pp. 549–571.

[22] Clarricoats, P. J. B., and Chan, K. B., (1973), Propagation Behaviour of Cylindrical Dielectric-Rod Waveguides, *Proceedings of the IEE*, Vol. 120, No. 11, pp. 1371–1378.

[23] Deschamps, G. A., (1971), Gaussian Beam as a Bundle of Complex Rays, *Electronics Letters*, Vol. 7, No. 23, pp.684-685.

Chapter 6

Dielectric Cone

6.1 INTRODUCTION

The first use of dielectric cones in microwave antennas appears to be due to Bartlett and Moseley [1] who patented the use of the "dielguide," a low-permittivity dielectric cone, placed between the feed and the subreflector of a reflector antenna. The dielectric cone played a dual role both as the feed support, reducing strut blockage, and as a "waveguide" linking the feed to the subreflector, to decrease spillover. The introduction of the "dielguide" implied modifying the subreflector shape from the traditional hyperboloid, using geometric optics.

Clarricoats [2] noted the similarities between the electromagnetic behavior of optical waveguides and corrugated feeds and suggested that the low-permittivity dielectric cone could perform as an antenna much like the corrugated horn.

Clarricoats and Salema in a series of papers [3, 4, 5, 6, 7, 8] put forward a modal theory for the propagation of electromagnetic waves in dielectric cones, calculated the excitation efficiency for the first few modes, predicted the radiation pattern of dielectric cones of very low to moderate permittivity, including the effect of the launcher, and suggested the use of an absorbent shield to reduce launcher influence in the radiation pattern, particularly in the far-out sidelobe region. They also presented design criteria for dielectric cone antennas on their own and as a subreflector support for reflector antennas.

Lier [9] proposed the use of a dielectric loaded metallic conical horn as a replacement for the corrugated horn.

Dielectric loaded metal conical horns were extensively researched by Olver [10] as a means of obtaining a high-performance feed with very low cross-polarization levels for reflector antennas particularly at millimeter wavelength where corrugated horns are difficult to build and expensive.

Besides dielectric loaded metal horns, exhaustive reviews on dielectric antennas [10, 11] do not include further references to solid dielectric cones on their own as antennas.

In this chapter we develop an approximate solution of Maxwell's equations for a dielectric cone of small to moderate semiflare angle at large distances from the apex. The fields inside the dielectric cone are expressed in terms of associated Legendre functions and the fields outside the cone are approximated in the manner of the fields of a dielectric cylinder.

On the basis of the characteristic equation, it is shown that the fields have a "quasi-modal" structure, which, in the above conditions, is only slowly dependent on the distance from the apex. The similarity between the electromagnetic behavior of the dielectric cone and the metallic corrugated horn is demonstrated and it is shown that, when the distance from the apex tends to infinity, the dielectric cone behaves exactly as a corrugated horn.

The orthogonality property is numerically demonstrated for the "quasi-modes" of the proposed solution and the excitation of the first quasi-modes is calculated using modal matching techniques.

Based on the transverse fields, the radiation pattern of a truncated cone is calculated using a vector Kirchhoff-Huygens integration over the spherical cap (centred on the apex) that constitutes the equiphase surface for each quasi-mode. Calculations show that this structure has excellent properties as a microwave antenna, providing an almost side-lobe-free circular symmetric pattern that, by proper design, can be made almost frequency-independent over a fairly large bandwidth.

The influence of the launching horn on the radiation pattern of the dielectric cone is numerically evaluated by a simple method and metallic shields are proposed as an effective means of reducing launcher-induced sidelobes in the resultant radiation pattern.

Although the material provided may be used in the design of a dielguide, particularly as regards field distribution and excitation efficiency, we do not deal with this subject here and refer the interested reader to [1, 8].

6.2 AN APPROXIMATE SOLUTION OF MAXWELL'S EQUATIONS

We shall look for a solution of Maxwell's equations for a dielectric cone starting with the more general (and in practice more important) case of hybrid modes. As will be shown below, these modes are not true modes, since they slowly change form as they propagate and should rather be referred to as quasi-modes. For simplicity however, they will henceforth be called modes. TE and TM modes will later be treated as special cases of the hybrid modal solution.

6.2.1 Hybrid Mode Characteristic Equation

Consider an infinite lossless dielectric cone immersed in an otherwise free space and a system of spherical coordinates r, θ, ϕ as depicted in Figure 6.1

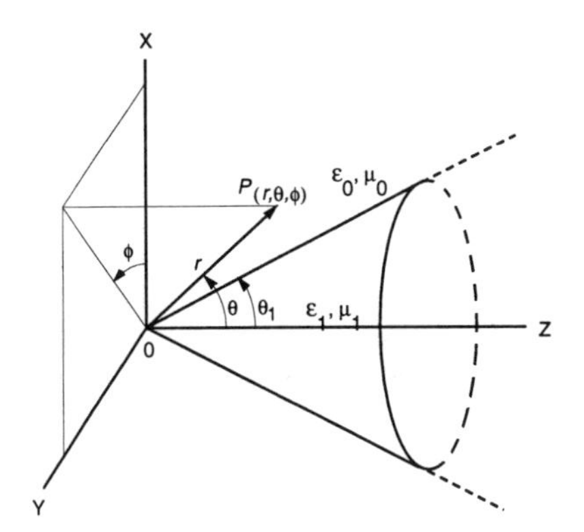

Figure 6.1: Coordinate system for the dielectric cone.

Taking for the fields an $\exp(j\omega t)$ dependence, the magnetic and electric vector potentials, $\mathcal{A}$ and $\mathcal{F}$, may be written as [12]

$$\mathcal{A} = \mu A_r \boldsymbol{r} \tag{6.1}$$

$$\mathcal{F} = \varepsilon F_r \boldsymbol{r} \tag{6.2}$$

where $\boldsymbol{r}$ is the unit vector in the r direction.

Assuming separation of variables and omitting a common factor $\exp(j\omega t)$ we have:

$$A_{r_1} = \sum_{n,\nu} a_{n,\nu} \hat{B}_\nu(k_1 r) P_\nu^n(\cos\theta)\sin(n\phi) \tag{6.3}$$

$$A_{r_0} = \sum_{n,\tau} b_{n,\tau} \hat{B}_\tau(k_0 r) G(\theta)\sin(n\phi) \tag{6.4}$$

$$F_{r_1} = \sum_{m,\nu} c_{n,\nu} \hat{B}_\nu(k_1 r) P_\nu^n(\cos\theta)\cos(n\phi) \tag{6.5}$$

$$F_{r_0} = \sum_{n,\tau} d_{n,\tau} \hat{B}_\tau(k_0 r) G(\theta)\cos(n\phi) \tag{6.6}$$

where $a_{n,\nu}$, $b_{n,\tau}$, and $C_{n,\nu}$, are arbitrary constants to be determined by matching the boundary conditions; $P_\nu^n(\cos\theta)$ is the associated Legendre function of the first kind, with an integer superscript n and a real subscript ν; $G(\theta)$ is a solution of the associated Legendre equation, to be specified later; and $\hat{B}_\nu$ and $\hat{B}_\tau$ are, respectively, solutions of spherical Bessel functions, used by Schelkunoff, which satisfy the following differential equations:

$$\left[\frac{d^2}{dr^2} + k_1^2 - \frac{\nu(\nu+1)}{r^2}\right]\hat{B}_\nu(k_1 r) = 0 \tag{6.7}$$

$$\left[\frac{d^2}{dr^2} + k_0^2 - \frac{\tau(\tau+1)}{r^2}\right]\hat{B}_\tau(k_0 r) = 0 \tag{6.8}$$

Subscripts 1 and 0 denote the regions inside and outside the cone, respectively, and, as usual,

$$k_1^2 = \omega^2 \varepsilon_1 \mu_1 \tag{6.9}$$
$$k_0^2 = \omega^2 \varepsilon_0 \mu_0 \tag{6.10}$$

The choice of circular functions for the ϕ dependence corresponds to a given polarization. Exchanging the sine and cosine functions in (6.3) to (6.6) results in a $\pi/2$ rotation of polarization.

Restricting ourselves to values such that $k_0 r \gg 1$ and considering only the outgoing wave, we will enforce a solution where

$$\hat{B}_\nu(k_1 r) \rightarrow \exp(-j\beta r) \tag{6.11}$$
$$\hat{B}_\tau(k_0 r) \rightarrow \exp(-j\beta r) \tag{6.12}$$

The former assumption, together with (6.7) and (6.8) leads to

$$\beta^2 = k_1^2 - \frac{\nu(\nu+1)}{r^2} \tag{6.13}$$
$$\beta^2 = k_0^2 - \frac{\tau(\tau+1)}{r^2} \tag{6.14}$$

From the magnetic and electric vector potentials $\mathcal{A}$ and $\mathcal{F}$ as defined in (6.1) and (6.2) the fields can be obtained using (2.61) to (2.64) in spherical coordinates. For convenience we write down this set of equations [12]:

$$E_r = \frac{1}{j\omega\varepsilon}\left(\frac{\partial^2}{\partial r^2} + k^2\right) A_r \tag{6.15}$$
$$E_\theta = -\frac{1}{r\sin(\theta)}\frac{\partial F_r}{\partial \phi} + \frac{1}{j\omega\varepsilon\, r}\frac{\partial^2 A_r}{\partial r \partial \theta} \tag{6.16}$$
$$E_\phi = \frac{1}{r}\frac{\partial F_r}{\partial \theta} + \frac{1}{j\omega\varepsilon\, r\sin(\theta)}\frac{\partial^2 A_r}{\partial r \partial \phi} \tag{6.17}$$
$$H_r = \frac{1}{j\omega\mu}\left(\frac{\partial^2}{\partial r^2} + k^2\right) F_r \tag{6.18}$$
$$H_\theta = \frac{1}{r\sin(\theta)}\frac{\partial A_r}{\partial \phi} + \frac{1}{j\omega\mu\, r}\frac{\partial^2 F_r}{\partial r \partial \theta} \tag{6.19}$$
$$H_\phi = -\frac{1}{r}\frac{\partial A_r}{\partial \theta} + \frac{1}{j\omega\mu\, r\sin(\theta)}\frac{\partial^2 F_r}{\partial r \partial \phi} \tag{6.20}$$

Substituting the values for the potential functions in the field expressions, taking into account (6.7) and (6.8) and dropping a common factor $\frac{\exp[j(\omega t - \beta r)]}{r}$, we get, for the "quasi-mode" n, ν, τ inside (subscript 1) and outside (subscript 0), the dielectric cone

$$E_{r_1} = -ja_{n,\nu}\frac{\nu(\nu+1)}{\omega\varepsilon_1 r}P_\nu^n(\cos\theta)\sin(n\phi) \tag{6.21}$$

$$E_{r_0} = -jb_{n,\tau}\frac{\tau(\tau+1)}{\omega\varepsilon_0 r}G(\theta)\sin(n\phi) \tag{6.22}$$

$$E_{\theta_1} = \left[-a_{n,\nu}\frac{\beta}{\omega\varepsilon_1}\frac{dP_\nu^n(\cos\theta)}{d\theta} + c_{n,\nu}n\frac{P_\nu^n(\cos\theta)}{\sin\theta}\right]\sin(n\phi) \tag{6.23}$$

$$E_{\theta_0} = \left[-b_{n,\tau}\frac{\beta}{\omega\varepsilon_0}\frac{dG(\theta)}{d\theta} + d_{n,\tau}n\frac{G(\theta)}{\sin\theta}\right]\sin(n\phi) \tag{6.24}$$

$$E_{\phi_1} = \left[-a_{n,\nu}\frac{n\beta}{\omega\varepsilon_1}\frac{P_\nu^n(\cos\theta)}{\sin\theta} + c_{n,\nu}\frac{dP_\nu^n(\cos\theta)}{d\theta}\right]\cos(n\phi) \tag{6.25}$$

$$E_{\phi_0} = \left[-b_{n,\tau}\frac{n\beta}{\omega\varepsilon_0}\frac{G(\theta)}{\sin\theta} + d_{n,\tau}\frac{dG(\theta)}{d\theta}\right]\cos(n\phi) \tag{6.26}$$

$$H_{r_1} = -jc_{n,\nu}\frac{\nu(\nu+1)}{\omega\mu_1 r}P_\nu^n(\cos\theta)\cos(n\phi) \tag{6.27}$$

$$H_{r_0} = -jd_{n,\tau}\frac{\tau(\tau+1)}{\omega\mu_0 r}G(\theta)\cos(n\phi) \tag{6.28}$$

$$H_{\theta_1} = \left[a_{n,\nu}n\frac{P_\nu^n(\cos\theta)}{\sin\theta} - c_{n,\nu}\frac{\beta}{\omega\mu_1}\frac{dP_\nu^n(\cos\theta)}{d\theta}\right]\cos(n\phi) \tag{6.29}$$

$$H_{\theta_0} = \left[b_{n,\tau}n\frac{G(\theta)}{\sin\theta} - d_{n,\tau}\frac{\beta}{\omega\mu_0}\frac{dG(\theta)}{d\theta}\right]\cos(n\phi) \tag{6.30}$$

$$H_{\phi_1} = \left[-a_{n,\nu}\frac{dP_\nu^n(\cos\theta)}{d\theta} + c_{n,\nu}\frac{n\beta}{\omega\mu_1}\frac{P_\nu^n(\cos\theta)}{\sin\theta}\right]\sin(n\phi) \tag{6.31}$$

$$H_{\phi_0} = \left[-b_{n,\tau}\frac{dG(\theta)}{d\theta} + d_{n,\tau}\frac{n\beta}{\omega\mu_0}\frac{G(\theta)}{\sin\theta}\right]\sin(n\phi) \tag{6.32}$$

Boundary conditions impose the continuity of E_r, H_r, E_ϕ, and H_ϕ for $\theta = \theta_1$. For E_r and H_r we get, respectively,

$$\frac{b_{n,\tau}}{a_{n,\nu}} = \frac{\varepsilon_0\nu(\nu+1)P_\nu^n(\cos\theta_1)}{\varepsilon_1\tau(\tau+1)G(\theta_1)} \tag{6.33}$$

$$\frac{d_{n,\tau}}{c_{n,\nu}} = \frac{\mu_0\nu(\nu+1)P_\nu^n(\cos\theta_1)}{\mu_1\tau(\tau+1)G(\theta_1)} \tag{6.34}$$

From the continuity of E_ϕ, taking into account (6.33) and (6.34), we get

$$\frac{a_{n,\nu}}{c_{n,\nu}} = \mp\frac{\omega\varepsilon_1}{n\beta}\frac{f_\nu^n(\cos\theta_1) - \frac{\mu_0\nu(\nu+1)}{\mu_1\tau(\tau+1)}g(\cos\theta_1)}{1 - \frac{\nu(\nu+1)}{\tau(\tau+1)}} \tag{6.35}$$

where

$$f_\nu^n(\cos\theta_1) = \frac{\sin\theta_1}{P_\nu^n(\cos\theta_1)}\left[\frac{dP_\nu^n(\cos\theta)}{d\theta}\right]_{\theta=\theta_1} \tag{6.36}$$

$$g(\cos\theta_1) = \frac{\sin\theta_1}{G(\theta_1)}\left[\frac{dG(\theta)}{d\theta}\right]_{\theta=\theta_1} \tag{6.37}$$

The continuity of H_ϕ, together with equations (6.33) and (6.34), yields

$$\frac{a_{n,\nu}}{c_{n,\nu}} = \mp\frac{n\beta}{\omega\mu_1}\frac{1 - \frac{\nu(\nu+1)}{\tau(\tau+1)}}{f_\nu^n(\cos\theta_1) - \frac{\varepsilon_0\nu(\nu+1)}{\varepsilon_1\tau(\tau+1)}g(\theta_1)} \tag{6.38}$$

The characteristic equation is now obtained from (6.35) and (6.38) as

$$\left[f_\nu^n(\cos\theta_1) - \frac{\varepsilon_0\nu(\nu+1)}{\varepsilon_1\tau(\tau+1)}g(\theta_1)\right]\left[f_\nu^n(\cos\theta_1) - \frac{\mu_0\nu(\nu+1)}{\mu_1\tau(\tau+1)}g(\theta_1)\right] = \frac{n^2\beta^2}{\omega^2\varepsilon_1\mu_1}\left[1 - \frac{\nu(\nu+1)}{\tau(\tau+1}\right]^2 \tag{6.39}$$

Making

$$\kappa_1^2 = \frac{\nu(\nu+1)}{r^2} \tag{6.40}$$

$$\kappa_0^2 = -\frac{\tau(\tau+1)}{r^2} \tag{6.41}$$

and noting that with the usual normalization we get from (6.13) and (6.14)

$$\bar{\kappa}_1^2 = \bar{\varepsilon}_1\bar{\mu}_1 - \bar{\beta}^2 \tag{6.42}$$

$$\bar{\kappa}_0^2 = \bar{\beta}^2 - 1 \tag{6.43}$$

the characteristic equation becomes

$$\left[\frac{\bar{\varepsilon}_1 f_\nu^n(\cos\theta_1)}{\bar{\kappa}_1^2} + \frac{g(\theta_1)}{\bar{\kappa}_0^2}\right]\left[\frac{\bar{\mu}_1 f_\nu^n(\cos\theta_1)}{\bar{\kappa}_1^2} + \frac{g(\theta_1)}{\bar{\kappa}_0^2}\right] = n^2\bar{\beta}^2\left(\frac{1}{\bar{\kappa}_1^2} + \frac{1}{\bar{\kappa}_0^2}\right)^2 \tag{6.44}$$

It can easily be shown that in this case the continuity of the tangential components of the electromagnetic fields at the dielectric cone boundary implies the correct behavior of normal field components on the same boundary.

To get a convenient form for $G(\theta)$, and subsequently for $g(\theta)$, let us reduce the characteristic equation (6.44) just obtained, to the case of the dielectric cylinder, by taking the limit when θ_1 tends to zero. From [13] it is

$$\begin{aligned}\lim_{\theta\to 0}[f_\nu^n(\cos\theta)] &= \lim_{\theta\to 0}\left[\frac{dP_\nu^n(\cos\theta)}{d\theta}\frac{1}{P_\nu^n(\cos\theta)}\right] \\ &= \frac{\sqrt{\nu(\nu+1)}\theta J_n'\left[\sqrt{\nu(\nu+1)}\theta\right]}{J_n\left[\sqrt{\nu(\nu+1)}\theta\right]}\end{aligned} \tag{6.45}$$

where J_n' denotes the derivative of the Bessel function of first kind and order n with respect to its argument.

Introducing (6.40) and noting that the radius of the equivalent dielectric cylinder is

$$\rho_1 = r\theta_1 \tag{6.46}$$

we have

$$\lim_{\theta_1\to 0}[f_\nu^m(\cos\theta_1)] = F_n(\kappa_1\rho_1) \tag{6.47}$$

where, as before, $F_n(x)$ is defined as

$$F_n(x) = x\frac{J_n'(x)}{J_n(x)} \tag{6.48}$$

If we choose for $G(\theta)$ the form $P_\tau^n(-\cos\theta)$, which is a solution of the associated Legendre equation, linearly independent from $P_\tau^n(\cos\theta)$ and continuous for all values of $\theta \neq 0$ and take the limit when θ tends to zero we have [13]

$$\lim_{\theta\to 0}\left[\frac{dP_\tau^n(-\cos\theta)}{d\theta}\frac{1}{P_\tau^n(-\cos\theta)}\right] = \frac{\sqrt{\tau(\tau+1)}\theta H_n^{(1)\prime}\left[\sqrt{\tau(\tau+1)}\theta\right]}{H_n^{(1)}\left[\sqrt{\tau(\tau+1)}\theta\right]} \tag{6.49}$$

where $H_n^{(1)}(x)$ denotes the Hankel function of the first kind, order n, and argument x, and $H_n^{(1)\prime}(x)$ its derivative with respect to the argument.

Since

$$K_n(x) = j^{n+1}\frac{\pi}{2}H_n^{(1)}(jx) \tag{6.50}$$

introducing (6.41) we have

$$\begin{aligned}\lim_{\theta_1 \to 0}[g(\theta_1)] &= \frac{\kappa_0\rho_1 K_n'(\kappa_0\rho_1)}{K_n(\kappa_0\rho_1)} \\ &= M(\theta_1)\end{aligned} \tag{6.51}$$

Assuming, as we did for the dielectric cylinder, that $\mu_1 = \mu_0$, the characteristic equation becomes

$$\left[\frac{\varepsilon_1}{\varepsilon_0}\frac{F_n(\kappa_1\rho_1)}{\kappa_1^2} + \frac{M_n(\kappa_0\rho_1)}{\kappa_0^2}\right]\left[\frac{F_n(\kappa_1\rho_1)}{\kappa_1^2} + \frac{M_n(\kappa_0\rho_1)}{\kappa_0^2}\right] = n^2\bar{\beta}^2\left(\frac{1}{\kappa_1^2} + \frac{1}{\kappa_0^2}\right)^2 \tag{6.52}$$

which is precisely the characteristic equation of a dielectric cylinder of radius ρ_1.

The form chosen for function $G(\theta)$ may give rise to computational difficulties because τ is complex and, more important, because for small values of θ, the series used to calculate $P_\tau^n(-\cos\theta)$ converges slowly. Recognizing that, for practical cases, most of the energy will be concentrated inside the dielectric cone or just outside it, we may approximate $G(\theta)$, in the all-important region immediately adjacent to the dielectric cone, as

$$G(\theta) = a\ K_n(\kappa_0\rho) \tag{6.53}$$

where a is a constant and ρ is defined in Figure 6.2.

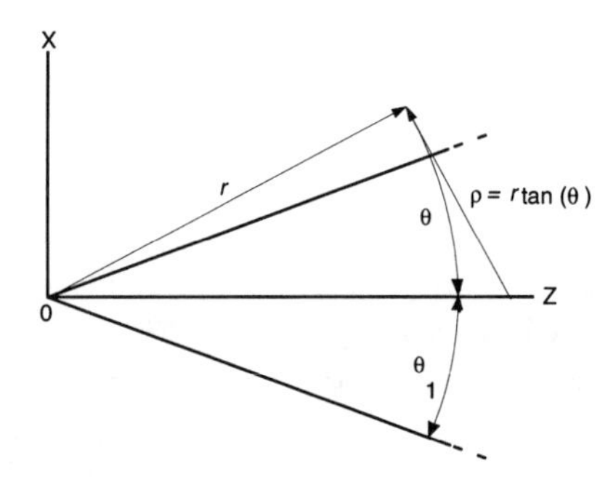

Figure 6.2: Definition of ρ in a dielectric cone.

It can be easily demonstrated that this new form for $G(\theta)$ yields exactly the same results for the characteristic equation in the limit when θ goes to zero.

Introducing (6.53) in (6.37) and (6.44), we finally obtain the characteristic equation for a dielectric cone at large distances from the apex:

$$\left[\frac{\bar{\varepsilon}_1 f_\nu^n(\cos\theta_1)}{\bar{\kappa}_1^2} + \frac{g(\theta_1)}{\bar{\kappa}_0^2}\right]\left[\frac{\bar{\mu}_1 f_\nu^n(\cos\theta_1)}{\bar{\kappa}_1^2} + \frac{g(\theta_1)}{\bar{\kappa}_0^2}\right] = n^2\bar{\beta}^2\left(\frac{1}{\bar{\kappa}_1^2} + \frac{1}{\bar{\kappa}_0^2}\right)^2 \tag{6.54}$$

where

$$g(\theta_1) = \frac{\kappa_0 r \tan(\theta_1) K_n'[\kappa_0 r \tan(\theta_1)]}{\cos(\theta_1) K_n[\kappa_0 r \tan(\theta_1)]} \tag{6.55}$$

The choice of $G(\theta)$ and the approximation used for the Schelkunoff's spherical Bessel functions have made the characteristic equation dependent on r thus violating the initial assumption of separation of variables. This is reflected in the fact that ν is not only a function of θ_1, $\bar{\varepsilon}_1$, and $\bar{\mu}_1$ but also of r. It can be numerically demonstrated, however, that, provided we restrict ourselves to large values of $k_0 r$, ν varies only slowly with r. Under such conditions we may confidently assume that the error entailed by using separation of variables is small.

If in (6.54) we take the limit, when r tends to infinity, we get

$$f_\nu^n(\theta_1) = \pm n \tag{6.56}$$

which is exactly the characteristic equation [14] for a corrugated horn operated under balanced hybrid conditions.

Numerically computed values of ν as a function of the cone semiflare angle θ_1 are represented in Figures 6.3 and 6.4, for the first hybrid mode of unity azimuthal dependence ($HE_{1,1}$) and different values of dielectric permittivity and normalized cone length.

As it will be shown below, the performance of a dielectric cone antenna can easily be predicted by comparison with the corrugated horn of the same slant length and the same value of ν. Hence, the solutions of the corrugated horn characteristic equation (6.56) in the $HE_{1,1}$ mode are also included in the previously mentioned figures.

Inspection of Figures 6.3 and 6.4 reveals that ν varies slowly with r showing the quasi-modal nature of our solution. Also the root of the characteristic equation of the corrugated horn is a close approximation of the root of the dielectric horn, particularly for the higher values of permittivity.

It is interesting to note that for many of cases of practical importance ($\theta_1 < 15$ degrees) the approximation in (6.45) gives almost exactly the same value of $\bar{\beta}$ (within 0.001 or better) as obtained using the associated Legendre functions. From the physical point of view, this is equivalent to saying that a dielectric cone behaves like a gradually expanding cylinder of radius $\rho_1 = r\theta_1$. The values of the fields in a cylinder can also be used as a good approximation for the fields in a dielectric cone, using $\rho = r\theta$.

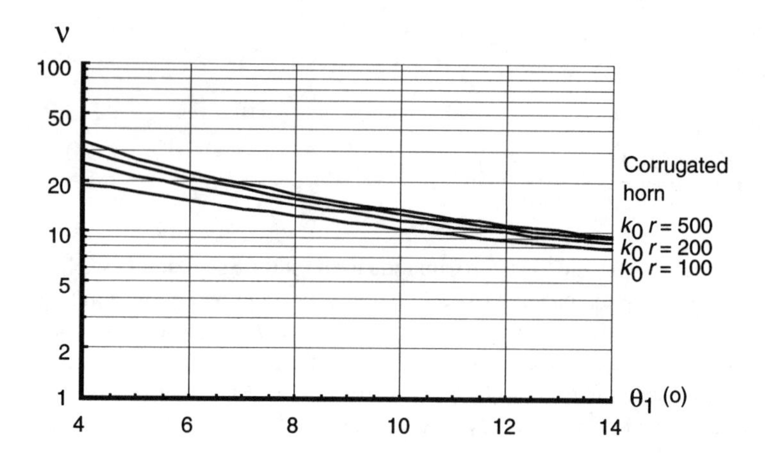

Figure 6.3: Root ν of the characteristic equation of a dielectric cone with $\bar{\varepsilon}_1 = 1.05$ versus cone semiflare angle θ_1 for the $HE_{1,1}$ mode. Also shown is the root of the characteristic equation of a corrugated horn for the $HE_{1,1}$ mode.

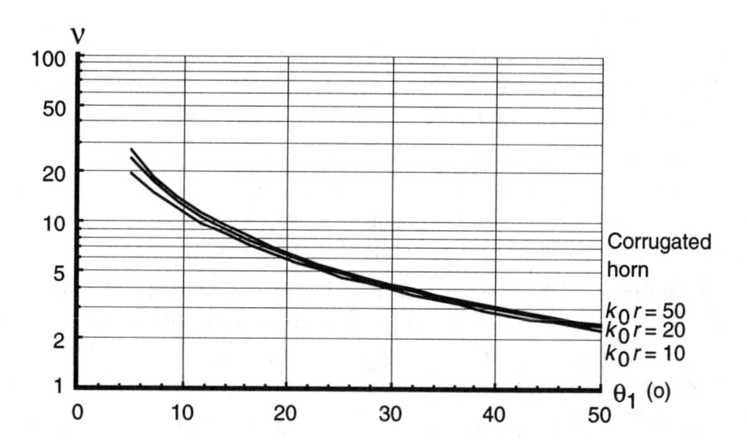

Figure 6.4: Root ν of the characteristic equation of a dielectric cone with $\bar{\varepsilon}_1 = 2.53$ versus cone semiflare angle θ_1 for the $HE_{1,1}$ mode. Also shown is the root of the characteristic equation of a corrugated horn for the $HE_{1,1}$ mode

However, it must be borne in mind that in the dielectric cone the fields propagate along the r direction and the equiphase surfaces are spheres, centered on the apex of the cone, whereas, in the cylinder, the equiphase surfaces are planes normal to the axis.

The orthogonality between modes of the different azimuthal dependence is easily demonstrated from the properties of the circular functions. For modes with the same azimuthal dependence, we could not demonstrate the orthogonality by analytical methods, and had to resort to numeric techniques. In this way we ascertain, for a number of different dielectric cones, that the modes as proposed are in fact orthogonal to each other.

6.2.2 Hybrid Mode Cutoff Frequency

As before we define cutoff frequency as the condition when the fields do not decay outside the dielectric cone. Below the cutoff frequency, the fields cannot be excited and therefore cannot exist.

At the cutoff point

$$\kappa_0 = 0 \tag{6.57}$$

which means that

$$\beta = k_0 \tag{6.58}$$

Considering that

$$\lim_{x\to 0}\left[\frac{x K_n{\prime}(x)}{K_n(x)}\right] = \infty \tag{6.59}$$

it is easily seen from (6.54) that the cutoff points are solutions of

$$P_\nu^n(\cos\theta_1) = 0 \tag{6.60}$$

For $n = 1$ the first zero of (6.60), for all values of θ_1, is zero. This means that the first hybrid mode of unity azimuthal dependence, the $HE_{1,1}$ mode, just like its counterpart in the dielectric cylinder, has no cutoff and may propagate irrespective of cone dimensions and permittivity. Of course, as the cone length, flare angle, or permittivity decrease, the mode fields extend further away from the dielectric and become increasingly difficult to excite.

The next five roots of (6.60) $\nu_{n,m}$, with $m = 2$–6, for modes with unity azimuthal dependence ($n = 1$), as a function of θ_1, are plotted in Figure 6.5. A good approximation

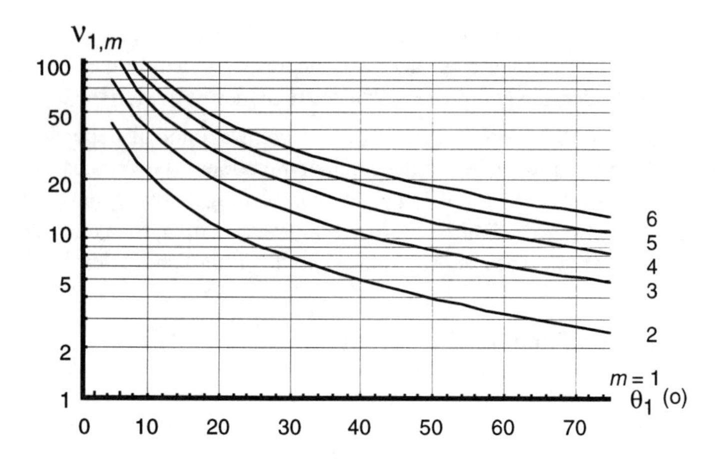

Figure 6.5: First six roots $\nu_{1,m}$ of $P_\nu^1(\cos\theta_1)$ versus θ_1.

for the values of $\nu_{n,m}$ may be obtained using the zeros of the Bessel function (first kind) that are to be found in tables such as Abramowitz and Stegun [15].

Noting that, at the cutoff points

$$\kappa_1^2 = \frac{\nu(\nu+1)}{r^2} \tag{6.61}$$

$$= k_0^2(\bar{\varepsilon}_1\bar{\mu}_1 - 1) \tag{6.62}$$

the cutoff condition can be expressed in terms of the minimum value of cone slant length r_{co}, below which a given mode cannot exist:

$$k_0 r_{co} = \sqrt{\frac{\nu_{n,m}(\nu_{n,m}+1)}{\bar{\varepsilon}_1\bar{\mu}_1 - 1}} \tag{6.63}$$

where $\nu_{n,m}$ denotes the m-order zero of (6.60).

According to the usual procedure for the dielectric cylinder, we will denote the hybrid modes, following the root order m as:

- $\mathrm{HE}_{n,l}$ for m odd;
- $\mathrm{EH}_{n,l}$ for m even;

where

$$l = (m-1) \div 2 + 1 \tag{6.64}$$

and the symbol $\div$ stands for integer division.

6.2.3 Hybrid Mode Fields

The hybrid mode fields are obtained from (6.21) to (6.32) taking into account the boundary conditions expressed in (6.33), (6.34),and (6.38).

Defining χ and ζ as

$$\chi = -\frac{\kappa_1^2 P_\nu^n(\cos\theta_1)}{\kappa_0^2 G(\theta_1)} \tag{6.65}$$

$$\zeta = \frac{f_\nu^n(\cos\theta_1) + \frac{\kappa_1^2}{\bar{\varepsilon}_1 \kappa_0^2} g(\theta_1)}{1 + \frac{\kappa_1^2}{\kappa_0^2}} \tag{6.66}$$

taking $\bar{\mu}_1 = 1$ and dropping an amplitude constant and the common factor $\frac{\exp[j(\omega t - \beta r)]}{r}$ we have

$$E_{r_1} = j\frac{\kappa_1^2 r}{\beta} P_\nu^n(\cos\theta)\sin(n\phi) \tag{6.67}$$

$$E_{r_0} = -j\frac{\kappa_0^2 r}{\beta} \chi G(\theta)\sin(n\phi) \tag{6.68}$$

$$E_{\theta_1} = \left[\frac{dP_\nu^n(\cos\theta)}{d\theta} - \frac{\bar{\varepsilon}_1\zeta}{\bar{\beta}^2}\frac{P_\nu^n(\cos\theta)}{\sin\theta}\right]\sin(n\phi) \tag{6.69}$$

$$E_{\theta_0} = \chi\left[\frac{dG(\theta)}{d\theta} - \frac{\bar{\varepsilon}_1\zeta}{\bar{\beta}^2}\frac{G(\theta)}{\sin\theta}\right]\sin(n\phi) \tag{6.70}$$

$$E_{\phi_1} = \left[n\frac{P_\nu^n(\cos\theta)}{\sin\theta} - \frac{\bar{\varepsilon}_1\zeta}{n\bar{\beta}^2}\frac{dP_\nu^n(\cos\theta)}{d\theta}\right]\cos(n\phi) \tag{6.71}$$

$$E_{\phi_0} = \chi\left[n\frac{G(\theta)}{\sin\theta} - \frac{\bar{\varepsilon}_1\zeta}{n\bar{\beta}^2}\frac{dG(\theta)}{d\theta}\right]\cos(n\phi) \tag{6.72}$$

$$H_{r_1} = -j\frac{\omega\varepsilon_1\kappa_1^2 r}{n\beta^2}\zeta P_\nu^n(\cos\theta)\sin(n\phi) \tag{6.73}$$

$$H_{r_0} = +j\frac{\omega\varepsilon_1\kappa_0^2 r}{n\beta^2}\chi\zeta G(\theta)\sin(n\phi) \tag{6.74}$$

$$H_{\theta_1} = -\frac{\omega\varepsilon_1}{\beta}\left[n\frac{P_\nu^n(\cos\theta)}{\sin\theta} - \frac{\zeta}{n}\frac{dP_\nu^n(\cos\theta)}{d\theta}\right]\cos(n\phi) \tag{6.75}$$

$$H_{\theta_0} = -\frac{\omega\varepsilon_1}{\beta}\frac{\chi}{\bar{\varepsilon}_1}\left[n\frac{G(\theta)}{\sin\theta} - \frac{\bar{\varepsilon}_1\zeta}{n}\frac{dG(\theta)}{d\theta}\right]\cos(n\phi) \tag{6.76}$$

$$H_{\phi_1} = \frac{\omega\varepsilon_1}{\beta}\left[\frac{dP_\nu^n(\cos\theta)}{d\theta} - \zeta\frac{P_\nu^n(\cos\theta)}{\sin\theta}\right]\sin(n\phi) \tag{6.77}$$

$$H_{\phi 0} = \frac{\omega \varepsilon_1}{\beta} \frac{\chi}{\bar{\varepsilon}_1} \left[\frac{dG(\theta)}{d\theta} - \bar{\varepsilon}_1 \zeta \frac{G(\theta)}{\sin\theta} \right] \sin(n\phi) \tag{6.78}$$

To simplify the interpretation of the previous equations we recall that

$$\frac{\omega \varepsilon_1}{\beta} = \frac{\bar{\varepsilon}_1}{\bar{\beta} Z_0} \tag{6.79}$$

where Z_0 is the characteristic free-space impedance in the outer medium.

The magnetic field, as a function of the angle from boresight, for the first five modes of unity azimuthal dependence ($n = 1$) is plotted in Figure 6.6 for a dielectric cone with $\bar{\varepsilon}_1 = 1.10$, $\theta_1 = 10$ degrees, and $k_0 r = 200$ and in Figure 6.7 for another dielectric cone with $\bar{\varepsilon}_1 = 2.53$, $\theta_1 = 30$ degrees, and $k_0 r = 30$. In these figures field amplitudes are normalized so that their maximum value for each mode is always unity.

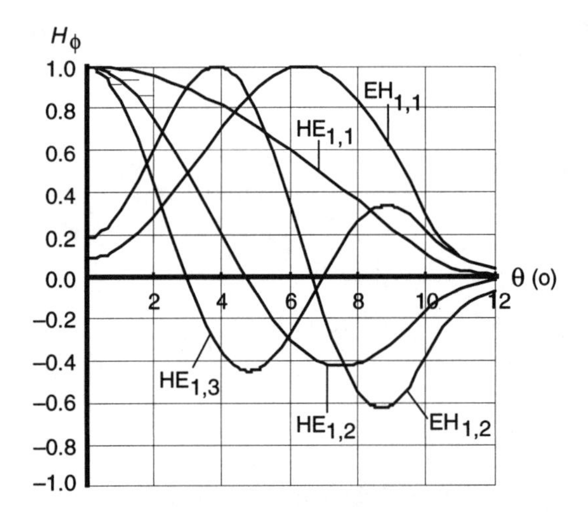

Figure 6.6: Aperture field pattern of the $HE_{1,1}$ mode of a dielectric cone with $\bar{\varepsilon}_1 = 1.10$, $\theta_1 = 10$ degrees, and $k_0 r = 200$.

These figures show clearly that the $HE_{1,m}$ mode fields always have a maximum on boresight while for the $EH_{1,m}$ modes this maximum is shifted away from boresight, with the shifting increasing as the mode gets further away from cutoff.

Although we focus our attention on the $HE_{1,1}$ mode, the effect of higher order modes must be taken into account to explain aperture field distributions, whenever dielectric cone parameters are such these modes can exist and the launcher is likely to excite them.

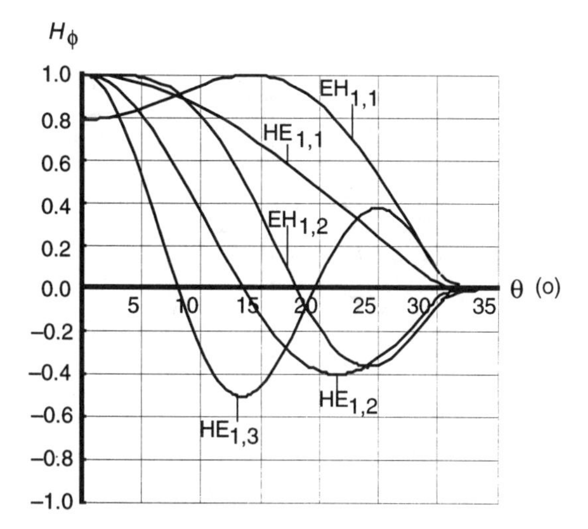

Figure 6.7: Aperture field pattern of the $HE_{1,1}$ mode of a dielectric cone with $\bar{\varepsilon}_1 = 2.53$, $\theta_1 = 30$ degrees and $k_0 r = 30$.

A typical example is given in Figure 6.8 for a dielectric cone, with $\bar{\varepsilon}_1 = 1.12, \theta_1 = 5.9$ degrees, and $k_0 r = 208.5$ excited by a conical metallic horn with $\theta_1 = 14$ degrees and normalized mouth radius $k_0 \rho = 14.9$, operated in the $TE_{1,1}$ mode. As shown, the $HE_{1,1}$ modal fields alone do not account very well for the measured aperture fields. However, when the next two modes ($EH_{1,1}$ and $HE_{1,2}$) are taken into account, the agreement between predicted and measured fields improves considerably. The relative amplitudes of the three modes were computed numerically using modal matching techniques as described in Section 3.6, and assuming no mode coupling between the launcher-dielectric cone interface and the dielectric cone mouth. The relative phases were also numerically computed at this interface. The phase differences, due to the propagation between the launcher and the aperture, were taken into account using the average of the values of $\bar{\beta}$ between the launcher and the aperture.

6.2.4 TM Modes

For the TM modes (6.21) to (6.32) apply with

$$c_c = d_c = 0 \tag{6.80}$$

The boundary conditions imply the continuity of E_r, E_ϕ, H_r, and H_ϕ. For E_r, (6.33) will still apply:

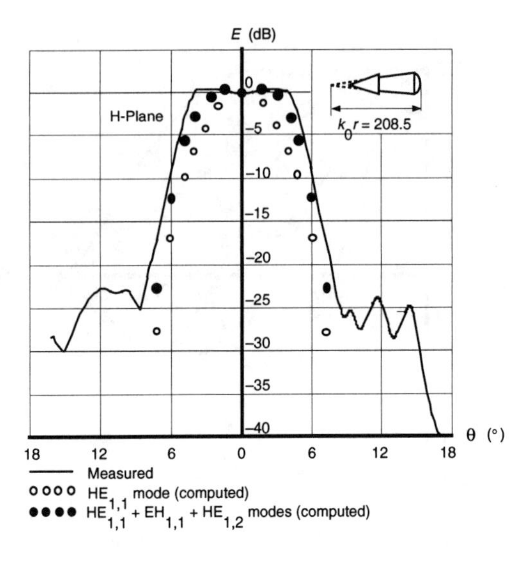

Figure 6.8: Aperture field pattern of the $HE_{1,1}$ mode of a dielectric cone excited by a conical metallic horn.

$$\frac{b_c}{a_c} = \frac{\varepsilon_0 \nu(\nu+1) P_\nu^n(\cos\theta_1)}{\varepsilon_1 \tau(\tau+1) G(\theta_1)} \tag{6.81}$$

The continuity of H_θ, together with (6.80) gives

$$\frac{b_c}{a_c} = \frac{\left[\frac{dP_\nu^n(\cos\theta)}{d\theta}\right]_{\theta=\theta_1}}{\left[\frac{dG(\theta)}{d\theta}\right]_{\theta=\theta_1}} \tag{6.82}$$

On the other hand, from (6.25), (6.26), (6.80), and (6.81), it is seen that the continuity of E_ϕ can only be ensured if $n = 0$. The characteristic equation is then

$$\frac{\bar{\varepsilon}_1 f_\nu^0(\cos\theta_1)}{\kappa_1^2} + \frac{g(\theta_1)}{\kappa_0^2} = 0 \tag{6.83}$$

The cutoff points, as defined before, are now given by

$$P_\nu^0(\cos\theta_1) = 0 \tag{6.84}$$

The roots of this equation are plotted in Figure 6.9.

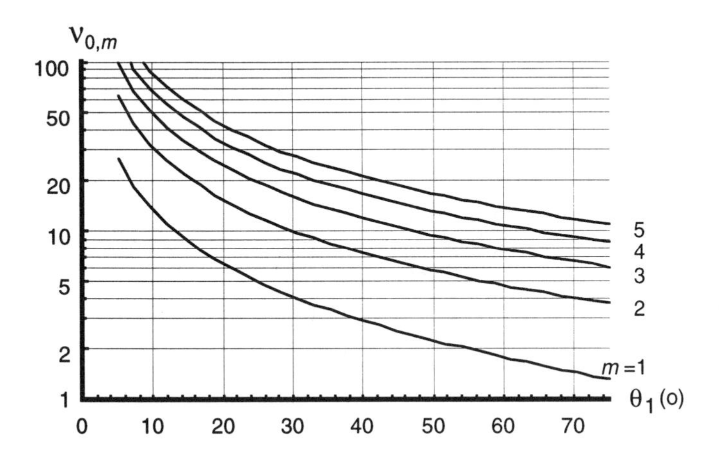

Figure 6.9: First five roots $\nu_{0,m}$ of $P_\nu^0(\cos\theta_1)$ versus θ_1.

Dropping an amplitude constant and the common factor $\frac{\exp[j(\omega t-\beta r)]}{r}$ the modal fields are

$$E_{r_1} = j\frac{\kappa_1^2 r}{\beta}P_\nu^0(\cos\theta) \tag{6.85}$$

$$E_{r_0} = -j\frac{\kappa_0^2 r}{\beta}\chi G(\theta) \tag{6.86}$$

$$E_{\theta_1} = \frac{dP_\nu^0(\cos\theta)}{d\theta} \tag{6.87}$$

$$E_{\theta_0} = \chi\frac{dG(\theta)}{d\theta} \tag{6.88}$$

$$E_{\phi_1} = E_{\phi_0} = 0 \tag{6.89}$$

$$H_{r_1} = H_{r_0} = 0 \tag{6.90}$$

$$H_{\theta_1} = H_{\theta_0} = 0 \tag{6.91}$$

$$H_{\phi_1} = \frac{\omega\varepsilon_1}{\beta}\frac{dP_\nu^0(\cos\theta)}{d\theta} \tag{6.92}$$

$$H_{\phi_0} = \frac{\omega\varepsilon_1}{\beta}\frac{\chi}{\bar{\varepsilon}_1}\frac{dG(\theta)}{d\theta} \tag{6.93}$$

where χ is defined in (6.65).

6.2.5 TE Modes

For the TE modes (6.21) to (6.32) will also apply with

$$a_c = b_c = 0 \tag{6.94}$$

The characteristic equation becomes

$$\frac{\bar{\mu}_1 f_\nu^n(\cos\theta_1)}{\kappa_1^2} + \frac{g(\theta_1)}{\kappa_0^2} = 0 \tag{6.95}$$

where $n = 0$ and the cutoff points are the same as those of the corresponding TM modes.

Similarly the modal fields dropping an amplitude constant and the common factor $\frac{\exp[j(\omega t - \beta r)]}{r}$ are

$$E_{r_1} = E_{r_0} = 0 \tag{6.96}$$

$$E_{\theta_1} = E_{\theta_0} = 0 \tag{6.97}$$

$$E_{\phi_1} = \frac{dP_\nu^0(\cos\theta)}{d\theta} \tag{6.98}$$

$$E_{\phi_0} = \frac{\chi}{\bar{\mu}_1}\frac{dG(\theta)}{d\theta} \tag{6.99}$$

$$H_{r_1} = j\frac{\kappa_1^2 r}{\omega\mu_1}P_\nu^0(\cos\theta) \tag{6.100}$$

$$H_{r_0} = -j\frac{\kappa_0^2 r}{\omega\mu_1}\chi G(\theta) \tag{6.101}$$

$$H_{\theta_1} = \frac{\beta}{\omega\mu_1}\frac{dP_\nu^0(\cos\theta)}{d\theta} \tag{6.102}$$

$$H_{\theta_0} = \frac{\beta}{\omega\mu_1}\chi\frac{dG(\theta)}{d\theta} \tag{6.103}$$

$$H_{\phi_1} = H_{\phi_0} = 0 \tag{6.104}$$

6.3 $HE_{1,1}$ MODE FIELDS

Let us examine in somewhat more detail the dominant $HE_{1,1}$ mode of the dielectric cone.

In Figures 6.10 and 6.11 we plot the normalized magnetic field H_ϕ of dielectric cones as a function of θ for varying slant lengths, semiflare angles, and dielectric permittivities. In Figure 6.10 we also plot the normalized magnetic field of the $HE_{1,1}$ balanced hybrid mode in a corrugated horn with the same semiflare angle as the dielectric cone, showing clearly that, as far a the field distribution is concerned, the horn behaves as an infinitely long dielectric cone.

If we take the lower permittivity ($\bar{\varepsilon}_1 = 1.05$) we see clearly that as the cone gets longer, the field concentrates more and more inside the dielectric. For the higher values of permittivity this behavior is accentuated and even for for moderate cone lengths ($k_0 r = 20$) the transverse fields are virtually nil outside the dielectric.

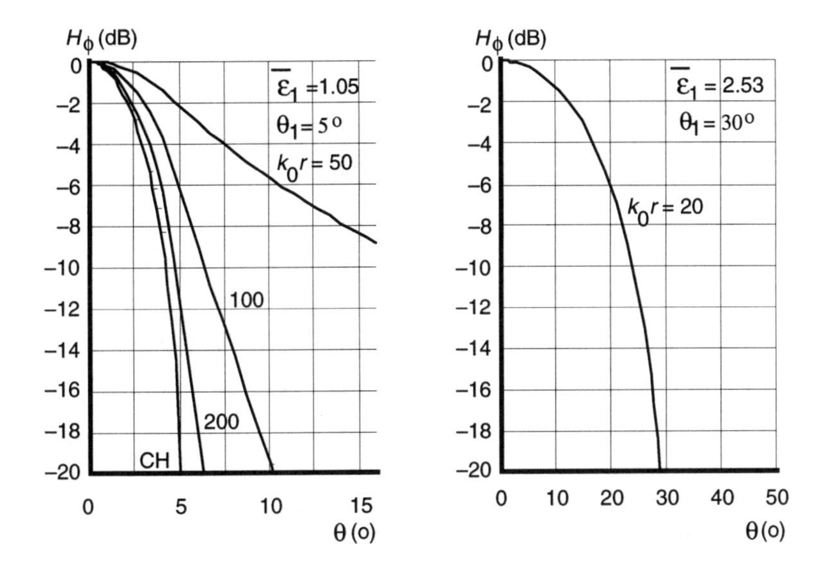

Figure 6.10: Calculated mouth field pattern of the $HE_{1,1}$ mode of a dielectric cone versus angle. Also shown is the mouth field pattern of a corrugated horn (CH) of the same semiflare angle.

The striking resemblance between the dielectric cone and the corrugated horn fields immediately suggests that the electromagnetic behavior of dielectric cones and corrugated horns should be similar. The similarity grows closer when the fields become more concentrated inside the cone, that is, at the same frequency for higher dielectric permittivities, larger semiflare angles, and longer cones.

As in the corrugated horn, the $HE_{1,1}$ mode fields in the dielectric cone are practically linearly polarized, especially for the lower dielectric constants ($\bar{\varepsilon}_1 \approx 1.05$) but also for moderately high dielectric constants ($\bar{\varepsilon}_1 \approx 2.5$).

Provided the dielectric permittivity is low, truncating the dielectric cone should not affect markedly the modal fields because their wave impedance is practically the same as free space. When the excitation of the higher-order modes is kept low (or if they are beyond "cutoff") and the excitation efficiency of the $HE_{1,1}$ mode is sufficiently high, the measured mouth field distributions are in excellent agreement with the theoretical results. Such is the case represented in Figure 6.12 where a dielectric cone with $\bar{\varepsilon}_1 = 1.12$, semiflare angle $\theta_1 = 6$ degrees, and normalized length $k_0 r = 150.8$, is excited by a corrugated waveguide, with a normalized mouth radius $k_0 \rho = 11.88$.

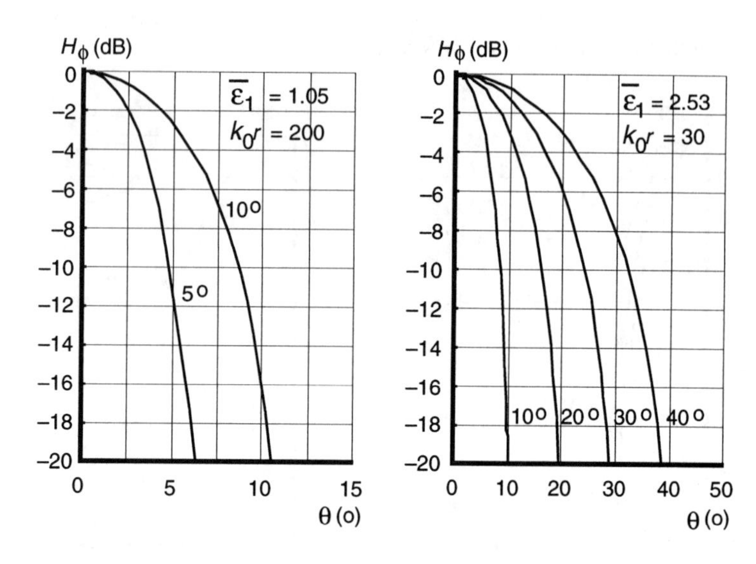

Figure 6.11: Calculated mouth field pattern of the $HE_{1,1}$ mode of a dielectric cone versus angle from axis for cone with different semiflare angles.

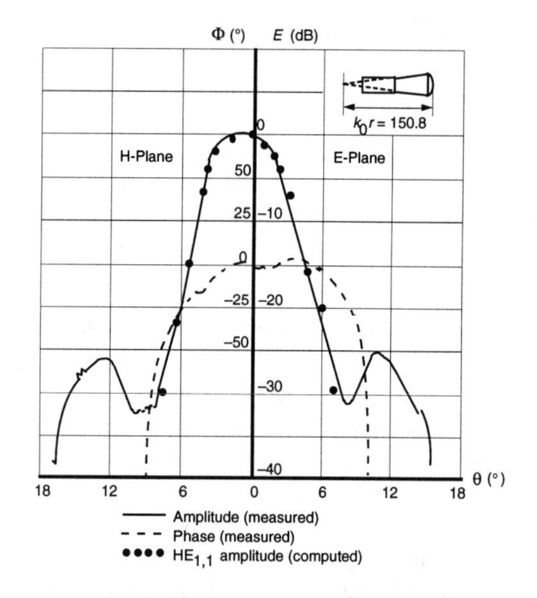

Figure 6.12: Mouth field distribution of the $HE_{1,1}$ mode of a dielectric cone ($\bar{\varepsilon}_1 = 1.12, \theta_1 =$ 6 degrees,and $k_0 r = 150.8$) excited by a corrugated waveguide ($k_0 \rho = 11.88$).

When the excitation efficiency is not as high as previously, the effect of the launcher becomes more pronounced and the agreement between the measured mouth fields and the calculated $HE_{1,1}$ mode fields is not as good. In Figure 6.13 we show the measured and calculated fields for the same dielectric cone ($\bar{\varepsilon}_1 = 1.045$, $\theta_1 = 6$ degrees, and $k_0 r = 150.8$), excited by a conical metallic horn, with $\theta_1 = 14$ degrees and normalized mouth radius $k_0 \rho = 11.97$ operated in the $TE_{1,1}$ mode.

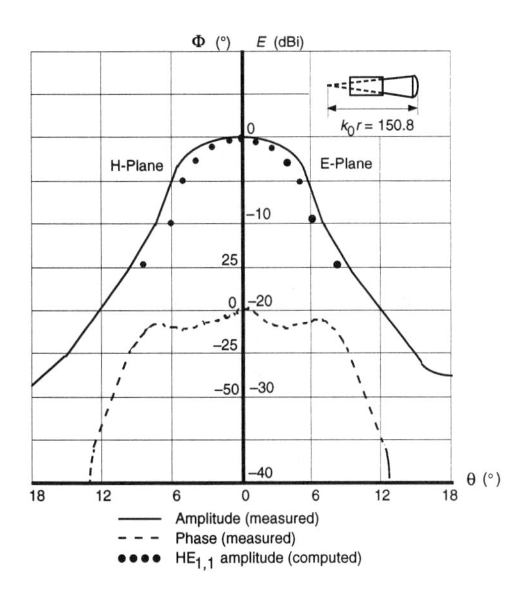

Figure 6.13: Mouth field distribution of the $HE_{1,1}$ mode of a dielectric cone ($\bar{\varepsilon}_1 = 1.045$, $\theta_1 = 6$ degrees,and $k_0 r = 150.8$) excited by a conical metallic horn ($\theta_1 = 14$ degrees, $k_0 \rho = 11.97$).

6.4 RADIATION PATTERN

The radiation pattern of a truncated dielectric cone may be obtained by a vector Kirchhoff-Huygens integration over the mouth and the side wall. In practical cases, where permittivity is low and cone dimensions are such that the fields concentrate inside the dielectric, we may safely neglect the contribution of the sidewall and perform the integration over the spherical cap that constitutes the equiphase surface. This cap should extend beyond the dielectric mouth to include the effect of modal fields outside the cone. As usual, the aperture fields are assumed to be the sum of the unperturbed incident fields plus a reflected wave due to the impedance mismatch between the dielectric cone modal fields and free space.

As will be shown below, a metallic shield may be added to the dielectric cone to suppress the influence of the launcher in the resulting radiation pattern. The shield is also very effective in attenuating any contribution of the sidewall to the radiation pattern. In that case, irrespective of permittivity, the radiation pattern may be computed integrating over the cone mouth, a spherical cap, centered on the cone apex and delimited by the shield.

6.4.1 $HE_{1,1}$ Mode

The computed far-field radiation patterns for the $HE_{1,1}$ mode of a set of low-permittivity dielectric cones are given in Figures 6.14 and 6.15. Since the radiation patterns are perfectly circular symmetric only the E-plane pattern is plotted. For convenience of comparison, all patterns are normalized to boresight, and the normalizing factor (boresight directivity, in dBi) is listed in Table 6.1.

Table 6.1: Boresight directivity, in dBi, for the set of low-permittivity dielectric cones whose E-plane radiation patterns are represented in Figures 6.14 and 6.15.

Dielectric constant $\bar{\varepsilon}_1$	Normalized length $k_0 r$	Semiflare angle $\theta_1 (^o)$	Directivity (dBi)
1.05	100	5	21.8
1.05	100	10	23.3
1.10	100	5	21.2
1.05	200	5	25.3

Figure 6.14, on the left, shows the effect of increasing the relative permittivity from 1.05 to 1.10 in a dielectric cone with $k_0 r = 100$ and semiflare angle $\theta_1 = 5$ degrees. As the fields concentrate inside the dielectric the effective aperture and the directivity decrease when the permittivity increases. The same figure on the right-hand side shows an apparently intriguing effect obtained when increasing the semiflare angle from 5 to 10 degrees in a dielectric cone with relative permittivity $\bar{\varepsilon}_1 = 1.05$ and $k_0 r = 100$. As the semiflare angle increases from 5° to 10°, the aperture increases but since the aperture fields tend to be more concentrated inside the dielectric, the net effect is quite small.

Figure 6.15 illustrates the effect of increasing the cone length. Here as the aperture size increases with the cone length the net result is a marked increase in directivity. The frequency dependence of the radiation pattern of a dielectric cone is equivalent to a change in cone length. Although not apparent from Figure 6.15, it is possible, by an appropriate choice of parameters, to design dielectric cones whose performance is virtually frequency-independent over a fairly large bandwidth.

As it could probably be expected from the quasi-Gaussian nature of the dielectric cone fields ($HE_{1,1}$ mode), the radiation pattern exhibits virtually no sidelobes. In addition, it

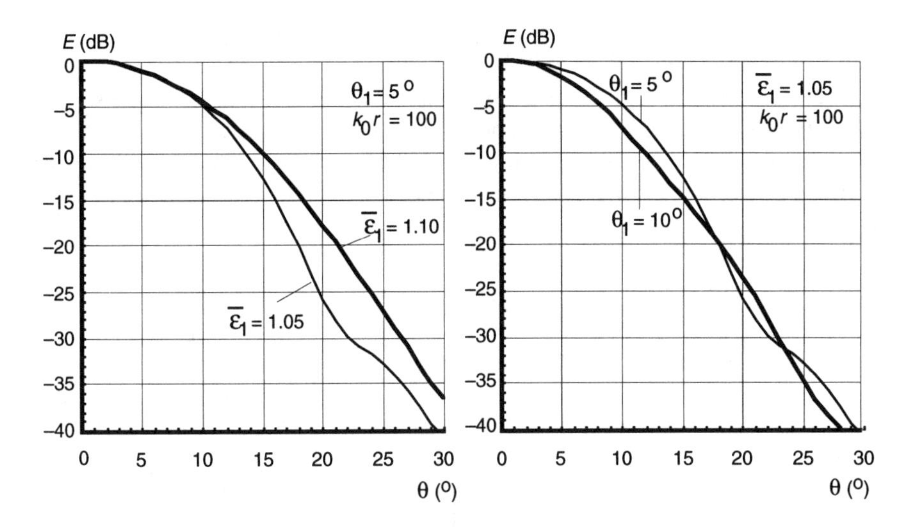

Figure 6.14: Computed E-plane radiation patterns of dielectric cones operated in the $HE_{1,1}$ mode.

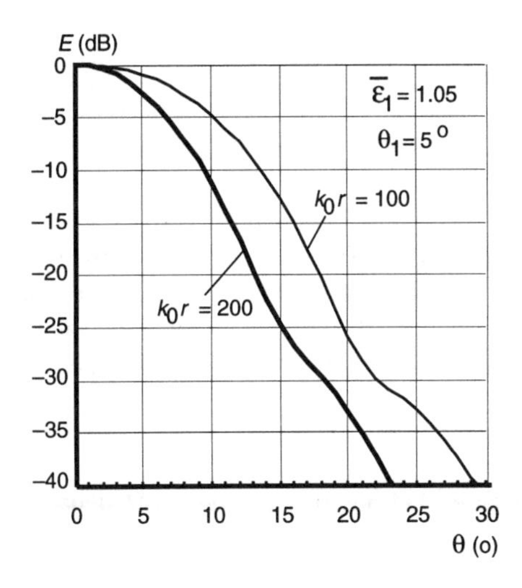

Figure 6.15: Computed E-plane radiation patterns of dielectric cones operated in the $HE_{1,1}$ mode.

has such a remarkable circular symmetry that there is no noticeable difference in the plots of the computed E- and H-plane radiation patterns.

In Figure 6.16 we plot the radiation patterns of the two dielectric cones operated in the $HE_{1,1}$ mode with $\bar{\varepsilon}_1 = 2.53$ and $k_0 r = 30$ and semiflare angles of 10 and 30 degrees, respectively. All patterns are normalized to boresight, and the normalizing factor (boresight directivity, in dBi) is listed in Table 6.2.

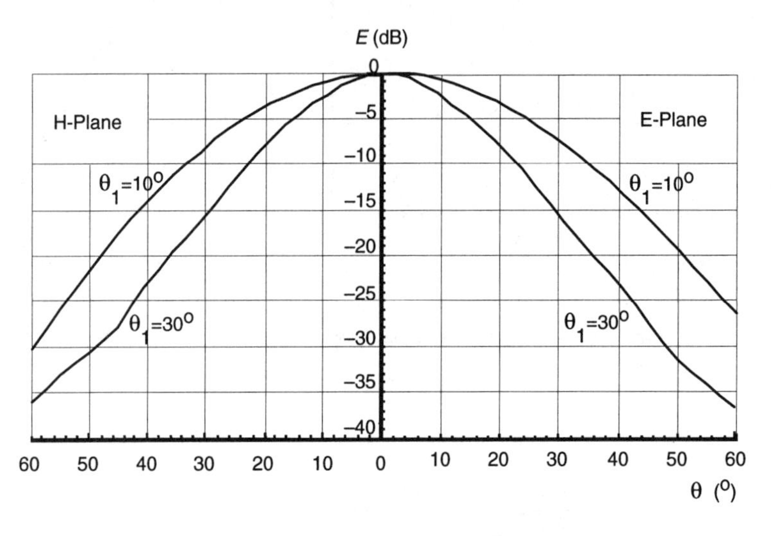

Figure 6.16: Computed far-field radiation patterns of dielectric cones with $\bar{\varepsilon}_1 = 2.53$ and $k_0 r = 30$ operated in the $HE_{1,1}$ mode.

The radiation patterns are almost, but not quite, circular symmetric and the directivity increases with the semiflare angle, up to a certain point. Although not shown in the figure the radiation pattern for a 20-degree semiflare angle cone is very similar to the one for the 30-degree cone. This is due to the phase error (for the equivalent plane aperture) which increases with the cone semiflare angle and which offsets the increase in directivity due to the larger aperture size.

6.4.2 Higher-Order Modes With Unity Azimuthal Dependence

In the corrugated horn the effect of modes other than the dominant $HE_{1,1}$ mode is generally not noticeable (except when these modes are deliberately excited) unless the cone semiflare angle is very large. In the dielectric cone, however, higher-order modes are usually excited and in some cases their presence contributes significantly to modify the aperture fields and the radiation pattern.

Table 6.2: Boresight directivity, in dBi, for the set of dielectric cones whose E-plane radiation patterns are represented in Figure 6.16.

Dielectric constant $\bar{\varepsilon}_1$	Normalized length $k_0 r$	Semiflare angle $\theta_1 (^o)$	Directivity D (dBi)
2.53	30	10	14.2
2.53	30	20	18.0
2.53	30	30	17.9

We represent in Figure 6.17 the computed E-plane and the H-plane radiation patterns (which are identical) of the $EH_{1,1}$ mode for a dielectric cone with $\bar{\varepsilon}_1 = 1.05$, $k_0 r = 200$, and $\theta_1 = 10$ degrees. Comparing these patterns with the $HE_{1,1}$ (Figure 6.15) the difference becomes obvious. Although both patterns are perfectly circular symmetric and have very low sidelobe levels, the $HE_{1,1}$ has a maximum on boresight while the $EH_{1,1}$ has a dip on boresight. The combination of both patterns with appropriate excitation efficiencies and phasing may be used to produce a "flat top" radiation pattern.

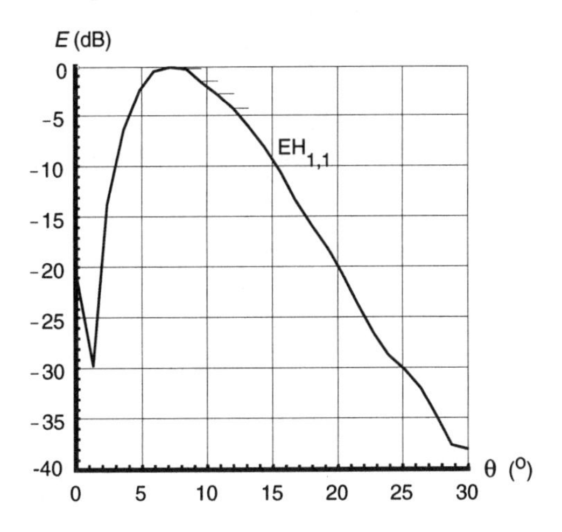

Figure 6.17: E-plane radiation pattern of a dielectric cone with $\bar{\varepsilon}_1 = 1.05$, $k_0 r = 200$, and $\theta_1 = 10$ degrees operated in the $EH_{1,1}$ mode.

Higher-order modes, such as the $HE_{1,2}$ and above, have high sidelobe levels and are thus usually avoided by an appropriate choice of the launcher that minimizes higher-order mode excitation efficiency (see Section 6.6) and/or by locating the excitation region where these modes are beyond cutoff.

The presence of higher-order modes in the radiation pattern of a dielectric cone has a number of practical consequences:

- Boresight directivity is usually slightly increased and the main lobe width decreased, due to increased illumination efficiency (apparent in Figure 6.8);
- Sidelobe level is increased;
- Performance becomes more frequency dependent because of the different mode velocities (for low-permittivity cones this effect is usually not too important, as these velocities are quite similar);
- The equiphase surface is no longer a sphere centered on the apex of the cone, due to the different mode velocities;
- The cross-polarization level may increase significantly.

6.4.3 Simplified Calculation Method

The Gaussian character of the $HE_{1,1}$ mode fields in a low-permittivity dielectric cylinder enabled us to suggest a simplified method to calculate the free-end radiation pattern. Under the conditions stated in Section 5.8 this method can be extended to dielectric cones of low semiflare angle, taking into account that the equiphase surface no longer as a plane but as a sphere of radius r.

Assume that the $HE_{1,1}$ field distribution in a dielectric cone of semiflare angle θ_1 is described by the same value of $\sqrt{2ak_0}$ as the dielectric cylinder of radius $\rho_1 = r\theta_1$. Taking

$$B(\rho_1) = \sqrt{2ak_0} \tag{6.105}$$

it is possible to define an equivalent dielectric cylinder whose aperture is located at a distance $R - z_0$ from the apex of the cone where

$$z_0 = \frac{r}{1 + \left(\frac{r}{a}\right)^2} \tag{6.106}$$

and is characterized by

$$a_0 = \frac{z_0 r}{a} \tag{6.107}$$

$$B_0 = \sqrt{2ak_0} \tag{6.108}$$

This equivalent dielectric cylinder has the same free-end radiation pattern as the original dielectric cone.

An alternative method, which leads to somewhat less accurate estimates, is to assimilate the dielectric cone mouth to a plane circular aperture with a cosine illumination and a quadratic phase error and then make use of expressions and graphs provided in Section 2.11.3. The maximum quadratic phase should be less than about π radians for reliable results.

Given a dielectric cone with normalized radius $\bar{r} = k_0 r$, semiflare angle θ_1 and relative dielectric permittivity $\bar{\varepsilon}_1$, the value of $\bar{\beta}$ may be estimated as for a dielectric cylinder with normalized radius $\bar{\rho}_1 = \bar{r}\theta_1$ using (5.44). The normalized radius of the equivalent plane circular aperture is given by

$$\bar{\rho}\text{eq} = \frac{\bar{\rho}_1}{1 - \frac{1}{1+\frac{\bar{\rho}_1}{2}\sqrt{\pi(\bar{\varepsilon}_1 - 1)}}} \tag{6.109}$$

The maximum phase error Φ, at the edge of the equivalent aperture, is

$$\Phi = \bar{\beta}\frac{\bar{\rho}^2_{\text{eq}}}{2\bar{r}} \tag{6.110}$$

Recalling (2.195) we may write the boresight directivity, di in dBi, as a function of $\bar{\rho}$eq:

$$di_{\text{dBi}} = 20\log_{10}(\bar{\rho}_{eq}) - 1.43 - \Delta_\Phi \tag{6.111}$$

where Δ_Φ is given as a function of Φ in Figure 2.17.

Finally the radiation pattern may be estimated using Figure 2.16. We must note that sidelobe levels obtained in this way tend to be considerably higher than the measured ones.

6.5 MAXIMUM CONE SEMIFLARE ANGLE

Although the theory developed so far does not place a well defined restriction on the value of the maximum semiflare angle, it can be easily confirmed experimentally that it cannot explain the behavior of low-permittivity ($\bar{\varepsilon}_1 < 1.10$), medium semiflare angle ($\theta_1 \approx 30$ degrees) dielectric cones, the presence of which hardly affects at all the radiation pattern of the launcher. It is nevertheless possible to explain this behavior from a simple geometric optics point of view.

Let us consider a dielectric cone in the receive mode (Figure 6.18). When the incoming rays meet the conical boundary at an angle θ_i less than the critical angle total internal reflection does not occur and the effect of the dielectric cone on the launcher is minimum. From this simple picture, it follows that the maximum semiflare angle is given by:

$$\theta_{1\,\max} = \frac{\pi}{2} - \theta_c$$

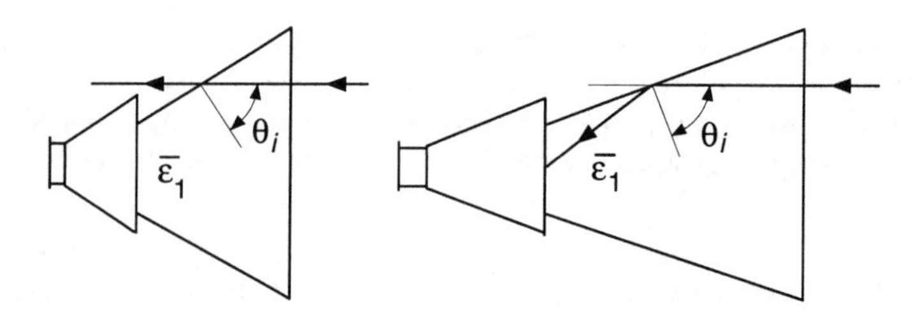

Figure 6.18: Diagram showing, under the geometric optics approach, the relation between maximum usable cone semiflare angle and dielectric permittivity.

$$= \frac{\pi}{2} - \arcsin\left(\frac{1}{\sqrt{\bar{\varepsilon}_1}}\right) \tag{6.112}$$

This is precisely the simplified concept of "trapping region" presented in Section 5.6.

With sufficiently high values of the dielectric permittivity, large semiflare angles may be used. The measured far-field radiation patterns for a dielectric cone, with semiflare angle $\theta_1 = 25$ degrees, and dielectric constant $\bar{\varepsilon}_1 = 2.26$, excited by a conical metallic horn, with normalized mouth radius $k_0\rho = 10.46$ and semiflare angle $\theta_1 = 24.2$ degrees are given in Figure 6.19. In the same figure we also represent the measured radiation pattern of the launcher and the computed radiation pattern for the $HE_{1,1}$ of the dielectric cone. The agreement between the theoretical and the measured results is very good on the H-plane, but not so good in the E-plane. The discrepancy can probably be accounted for by coupling between the launcher mouth and the dielectric cone aperture, as there was only a short length of cone protruding from the launcher.

The improvement in the radiation pattern of the launcher, as regards sidelobe and pattern symmetry, is well demonstrated in Figure 6.19.

The effect of frequency on the measured patterns is shown in Figure 6.20.

When relatively high values of dielectric permittivity are used, one should expect correspondingly large values of voltage standing wave ratio (VSWR). For the case referred to above, a maximum value of VSWR of 2.3 was measured at the X band.

As a first approximation the VSWR values can be calculated using the equivalent circuit in Figure 6.21. The maximum value of the voltage reflection coefficient Γ_{max}, due to the combined effects of the transitions between the launcher and the dielectric cone and the dielectric cone and free space, is given by

$$\Gamma_{max} = 2\frac{Z_1 - Z_0}{Z_1 + Z_0} \tag{6.113}$$

Since

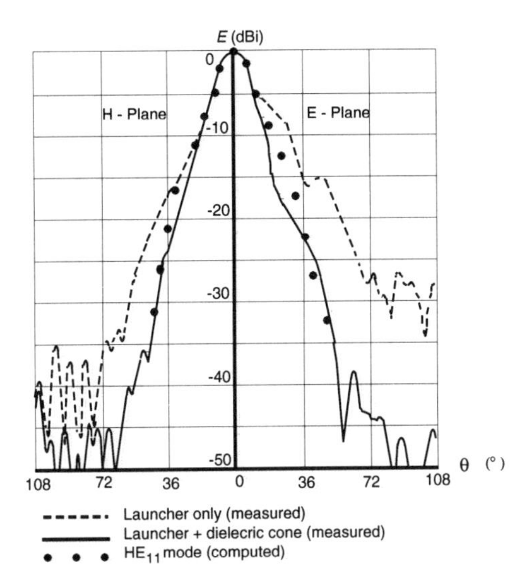

Figure 6.19: Far-field radiation pattern of a dielectric cone, with $\bar{\varepsilon}_1 = 2.26, \theta_1 = 25$ degrees and $k_0 r =$, excited by a conical metallic horn, with $\theta_1 = 24.2$ degrees and $k_0 \rho = 10.46$.

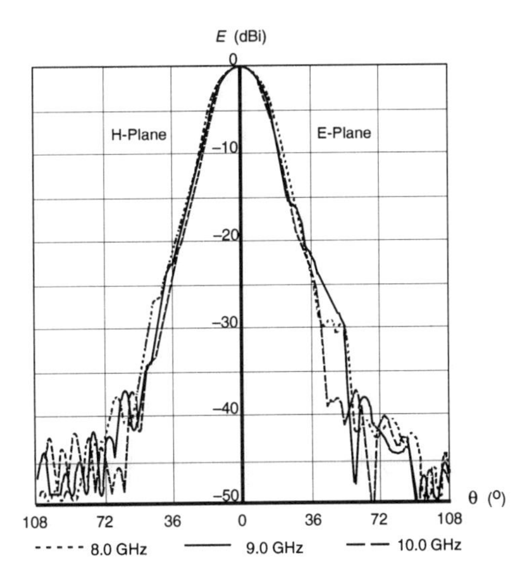

Figure 6.20: Effect of frequency on the measured far-field radiation pattern of the dielectric cone shown in Figure 6.19.

$$Z_1 \approx \frac{Z_0}{\sqrt{\bar{\varepsilon}_1}} \tag{6.114}$$

we get

$$\Gamma\text{max} = 2\frac{1-\sqrt{\bar{\varepsilon}_1}}{1+\sqrt{\bar{\varepsilon}_1}} \tag{6.115}$$

Making use of the relation between VSWR and voltage reflection coefficient:

$$VSWR\text{max} = \frac{1+|\Gamma\text{max}|}{1-|\Gamma\text{max}|} \tag{6.116}$$

we finally get

$$VSWR\text{max} = \frac{3\sqrt{\bar{\varepsilon}_1}-1}{3-\sqrt{\bar{\varepsilon}_1}} \tag{6.117}$$

Taking $\bar{\varepsilon}_1 = 2.26$, we obtain a maximum value for the VSWR of 2.35 that agrees closely with the measured value. Incidentally, this also explains why for values of dielectric permittivities up to 1.10 the measured values of VSWR never exceed about 1.10.

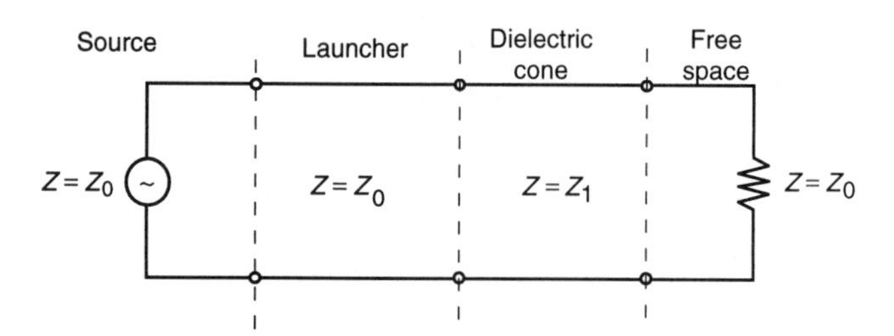

Figure 6.21: Equivalent circuit to calculate the maximum value of the VSWR for a dielectric cone antenna.

The VSWR can be improved, at least for a moderate frequency bandwidth, by matching the launcher-dielectric cone and the dielectric cone free space interfaces. A quarter-wavelength transformer may be used. This can be readily incorporated in the cone permittivity. One way of achieving this objective is to drill holes of small diameter and $\lambda_0/4$ deep normal to the incident wave front at the dielectric interfaces, in such a way that the dielectric density, and thus its permittivity, is reduced by the required amount.

6.6 EXCITATION EFFICIENCY OF DIELECTRIC CONE MODES

The excitation efficiency of the hybrid modes of a dielectric cone may be calculated by the modal matching technique described in Section 3.6, where the matching surface is a sphere centered on the apex of the dielectric cone, and spherical coordinates are used throughout.

To start with we will assume no phase error between the exciting field and the dielectric mode fields. As an example, we plot in Figure 6.22 the calculated efficiency for the excitation of the first five modes of a dielectric cone with $\theta_1 = 10$ degrees, $\bar{\varepsilon}_1 = 1.10$ by a metallic conical horn with $k_0 r = 200$, and varying semiflare angle θ_2 operated in the $TE_{1,1}$ mode. We should note that this example is not meant to represent a useful antenna (due to the launcher size) but rather to provide information on the excitation efficiencies of higher order modes. A metallic conical horn with $k_0 r \approx 50$ and $\theta_2 \approx 12.7$ degrees would provide a much more realistic launcher. In addition, in this case all higher-order modes would be beyond cutoff.

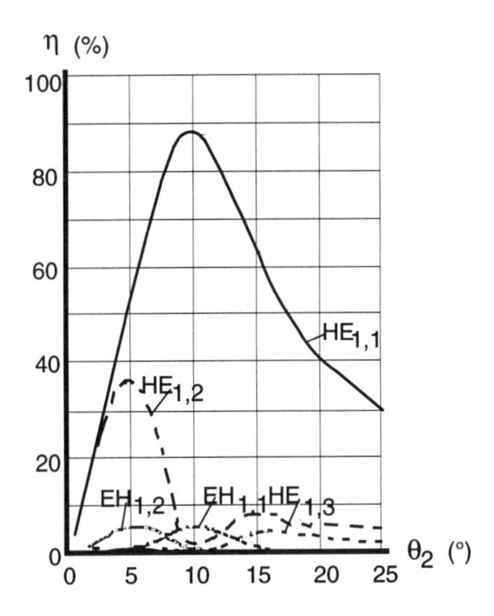

Figure 6.22: Excitation efficiency of the first five modes of unity azimuthal dependence of a dielectric cone with $\theta_1 = 10^o$, $\bar{\varepsilon}_1 = 1.10$, and $k_0 r = 200$, using as the launcher a conical metallic horn of varying semiflare θ_2, operated in the $TE_{1,1}$ mode.

Next we will consider a parabolic phase error with maximum value $\Delta\phi_{max}$ between the exciting field and the dielectric mode fields. Figure 6.23 depicts the effects of such phase errors in the excitation efficiency of the $HE_{1,1}$ mode of a dielectric cone by a conical horn operated in the $TE_{1,1}$. As shown, parabolic phase errors reduce the excitation efficiency but this reduction may be neglected, provided the maximum phase error does not exceed about 45 degrees.

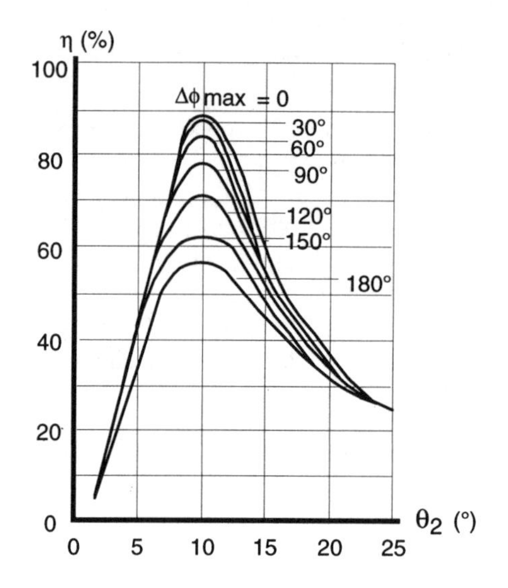

Figure 6.23: Effect of parabolic phase errors on the excitation efficiency of the $HE_{1,1}$ mode of a dielectric cone with $\theta_1 = 10$ degrees, $\bar{\varepsilon}_1 = 1.10$, and $k_0 r = 200$, using as the launcher a conical metallic horn of varying semiflare θ_2, operated in the $TE_{1,1}$ mode.

It can be shown that the use of a dielectric cylinder of radius $\rho_2 = r\theta_2$ and a circular waveguide of radius $\rho_2 = r\theta_2$ to calculate the excitation efficiency of the $HE_{1,1}$ mode gives accurate results for $\theta_1,\ \theta_2 < 20$ degrees. Under these conditions the conclusions drawn in Section 5.5 can still be applied.

6.7 EFFECT OF LAUNCHER

When a dielectric cone supporting the $HE_{1,1}$ dominant mode (or for that matter any set of cone modes) is used as an antenna, the observed radiation pattern is due not only to the cone but also to the launcher.

For low-permittivity dielectric materials, the method described in Section 5.8 and summarized in Figure 6.24 can be used to predict the effect of the launcher on the overall radiation pattern, noting that the trapping region is now limited by the sum of the complementary of the critical angle and the cone semiflare angle:

$$\theta_T = \frac{\pi}{2} - \arcsin\left(\frac{1}{\sqrt{\bar{\varepsilon}_1}}\right) + \theta_1 \tag{6.118}$$

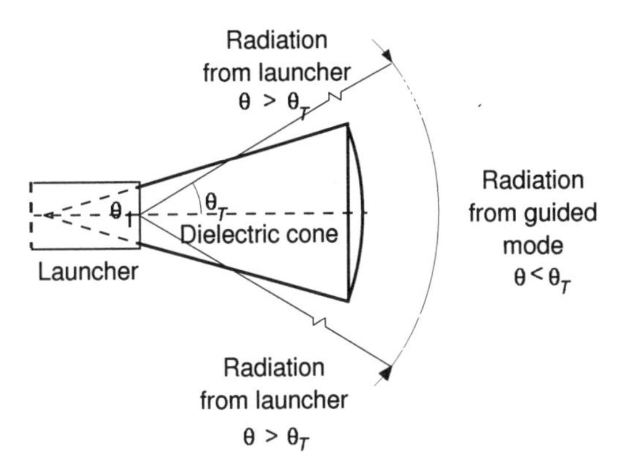

Figure 6.24: Diagram describing the method used to include the effect of the launcher in the radiation pattern of a low-permittivity dielectric cone.

The detailed calculation of the resulting pattern is exemplified in Figures 6.25 and 6.26 for the case of a pyramidal horn exciting a dielectric cone. As shown in these figures the agreement between the predicted and measured results is very good.

Experiments, performed with a dielectric cone with $\bar{\varepsilon}_1 = 1.12$, $\theta_1 = 6$ degrees, and $k_0 r = 150.8$ excited by a number of different available launchers, give further support to the method. For all these cases the computed patterns for the dielectric cone assume that only the $HE_{1,1}$ mode is present, because all other modes, according to our previous definition, are beyond cutoff. Although none of these launchers was optimized, it is obvious from the measured patterns that the resultant pattern has increased gain and improved circular symmetry. Also, because of the overall increase in gain and virtually sidelobe-free radiation pattern of the $HE_{1,1}$ mode, the relative level of the sidelobes of the resultant antenna is lower than that of launcher alone. Very simply, the above-mentioned patterns demonstrate that a considerable improvement in the launcher radiation pattern can be achieved by placing a low-permittivity dielectric cone in front of the launcher.

The use of a corrugated horn to excite a dielectric cone may at first seem a paradox, as the field patterns are so similar in both cases. However, the use of such a launcher may well be justified if one realizes that:

- The corrugated horn needs only be of sufficient size to obtain a reasonable excitation efficiency of the $HE_{1,1}$ mode in the dielectric cone;
- The length needed to obtain a given aperture size is achieved by the dielectric cone.

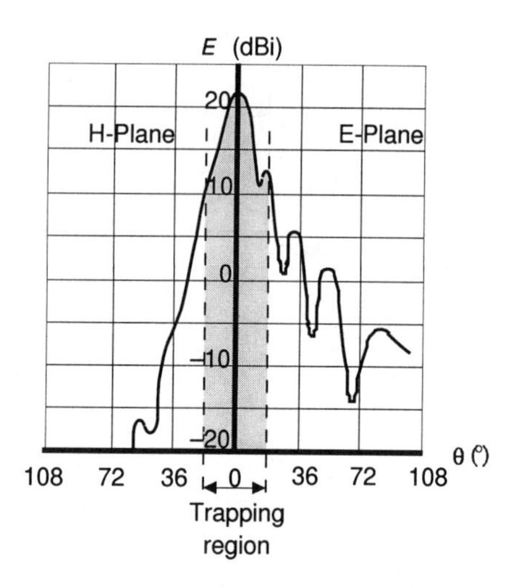

Figure 6.25: Calculated far-field radiation pattern of a 22-dBi standard-gain pyramidal horn (used as launcher) at 9 GHz, showing the trapping region for a dielectric cone with $\theta_1 = 6$ degrees,and $\bar{\varepsilon}_1 = 1.05$.

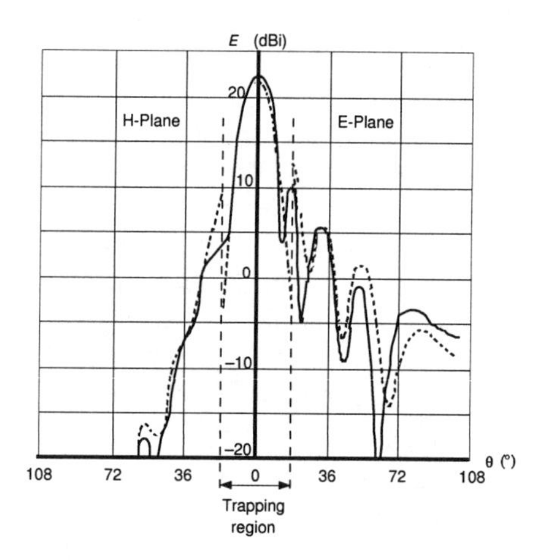

Figure 6.26: Far-field radiation pattern of a dielectric cone with $\bar{\varepsilon}_1 = 1.12$, $\theta_1 = 6$ degrees, and $k_0 r = 150.8$ excited by a pyramidal 22-dBi standard-gain horn at 9 GHz.

Experimental results are highly encouraging, as shown in Figures 6.27 and 6.28 where we represent the result radiation pattern and its frequency dependence compared with the launcher horn alone.

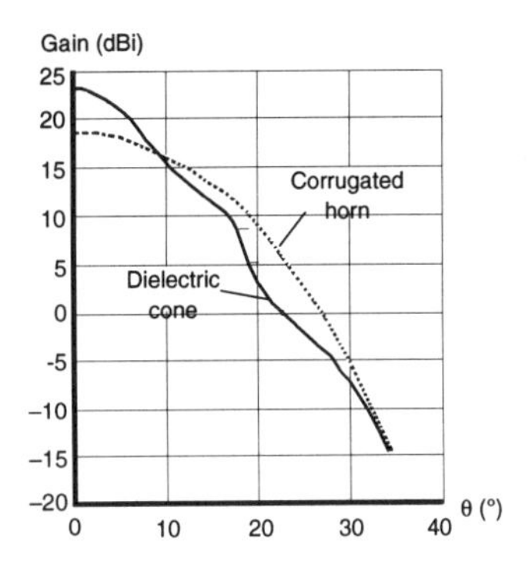

Figure 6.27: Measured E-plane far-field radiation patterns at 7.3 GHz of (1) a dielectric cone with $\bar{\varepsilon}_1 = 1.045$, $\theta_1 = 6$ degrees,and $k_0 r = 171.6$ excited by a corrugated horn, with $\theta_1 = 6$ degrees and normalized mouth radius $k_0\rho_1 = 9.63$ operated in the $HE_{1,1}$ mode; and (2) of the corrugated horn alone.

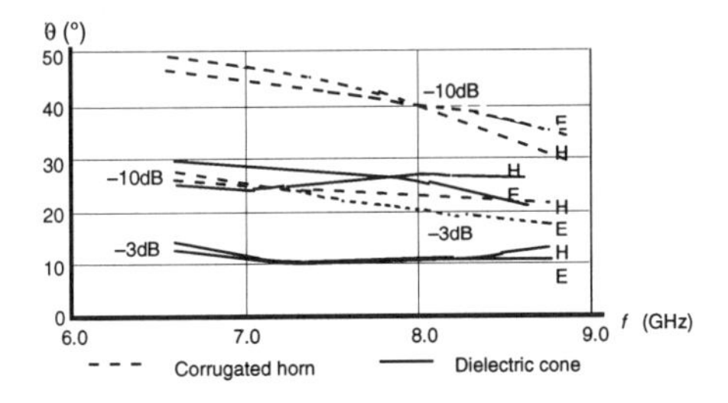

Figure 6.28: Measured beamwidths of the dielectric cone excited by the corrugated cone referred to in Figure 6.27 when the frequency varies from 6.6 to 8.8 GHz.

6.8 SUPPRESSING SIDEWALL AND LAUNCHER EFFECTS

In order to suppress radiation from the launcher Clarricoats and Salema [4] have suggested the use of a shield of absorbent material laid around the dielectric cone, in the same way as the absorbent coating of an optical fiber (Figure 6.29). As the energy is tightly bounded to the cone, the presence of the absorbent material has a negligible effect on the main lobe of the radiation pattern but, on the other hand, it is highly effective in suppressing the radiation from the launcher which occurs at angles greater than about θ_S.

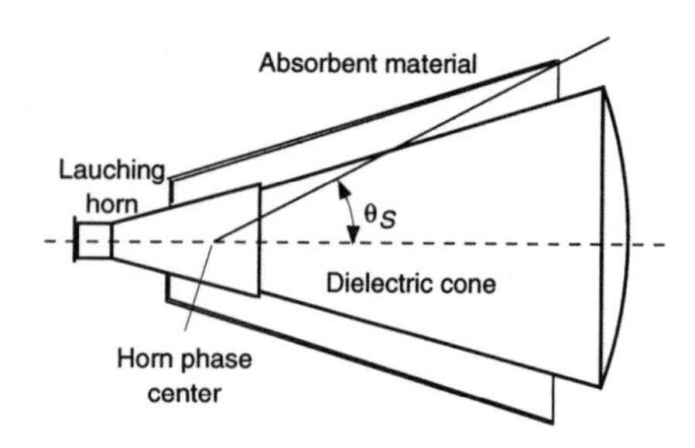

Figure 6.29: Diagram showing the use of an absorbent shield to suppress the effect of the launcher in the radiation pattern of a dielectric cone antenna.

Using a dielectric cone, with $\bar{\varepsilon_1} = 1.12$, $\theta_1 = 6$ degrees,and $k_0 r = 150.8$, excited by a conical metallic horn, with normalized aperture radius $k_0 \rho = 11.97$ and semiflare angle $\theta_1 = 14$ degrees and a layer of absorbent material such that $\theta_S \approx 15$ degrees, a significant reduction of the sidelobe levels was measured as indicated in Figure 6.30. Similar results were obtained using other types of launchers.

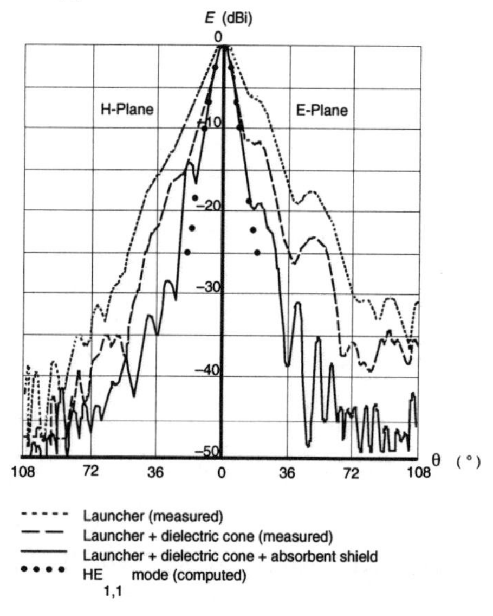

Figure 6.30: Effect of an absorbent shield on the measured far-field radiation pattern of a dielectric cone excited by a conical metallic horn operated in the $TE_{1,1}$ mode.

Carrying this idea forward, we can substitute the absorbent shield by a metallic one. If this shield shape and dimensions are such that the cone modal fields on it are negligible (in practice below about −20 dB from maximum value), the effect of the shield on the propagating modes can be neglected. However, the shield is effective in blocking the radiation from the launcher and from the sidewalls leaving only the desired free-end radiation pattern.

Figure 6.31 shows the computed and measured radiation pattern of a dielectric cone cone with relative permittivity $\bar{\varepsilon}_1 = 2.53$, semiflare angle $\theta_1 = 14.1$ degrees, and $k_0 r = 33.87$ enclosed in a metallic shield provided by a conical horn with a semiflare angle of 18.9 degrees. For simplicity the dielectric cone has a plane mouth rather than spherical cap one. This introduces a quadratic phase error on the aperture (assumed to be a spherical cap) whose maximum value is given by (6.110) where $\bar{\rho}_{eq}$ is now defined by the extent of the aperture. In the case shown in Figure 6.31 the maximum phase error was $\Phi = 0.569$ radians at $\theta = \theta_1$. The resulting radiation pattern exhibits very good circular symmetry up to -20 dB below boresight and low sidelobes.

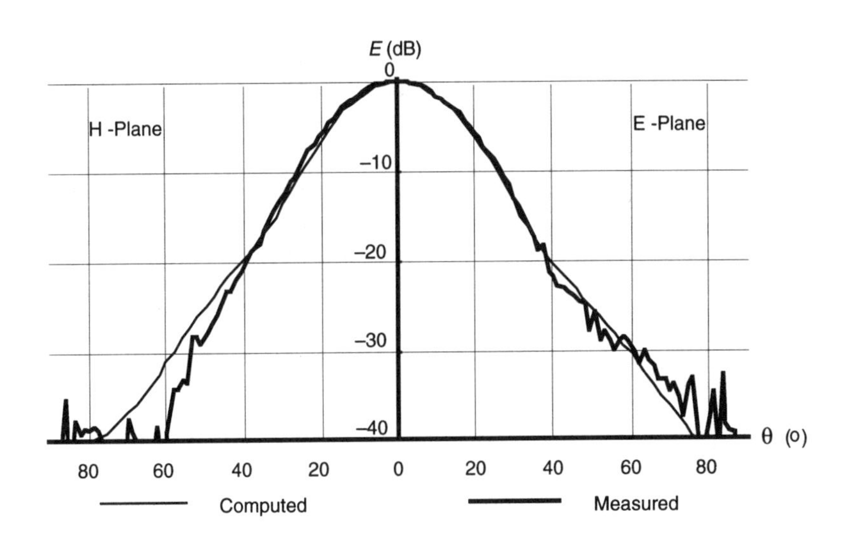

Figure 6.31: Computed and measured far-field radiation pattern of a shielded dielectric cone with relative permittivity $\bar{\varepsilon}_1 = 2.53$, semiflare angle $\theta_1 = 14.1$ degrees, and $k_0 r = 33.87$.

The dielectric cone is usually designed to operate in the $HE_{1,1}$ mode. Assuming the launcher to be a circular metallic waveguide at the transition, we must have a normalized cone radius less than the value given in (6.63), with $n = 1$ and $m = 2$, for the $EH_{1,1}$ mode to be below cutoff. As for the shielded dielectric pyramidal horn, this limit may be used to derive a maximum cone semiflare angle. In the example above a very simple mode filter helps in keeping down the amplitude of the $EH_{1,1}$ mode.

6.9 CONCLUSIONS

The dielectric cone may be used both as an antenna on its own and as a guiding structure between antenna elements, such as the feed and the subreflector in a Cassegrain reflector. The fundamental ($HE_{1,1}$) hybrid mode has a quasi-Gaussian field distribution with maximum value on boresight, low cross-polarization, and almost perfect circular symmetry. Field intensities on the lateral walls tend to be low but, for low values of permittivity and smaller cross-section dimensions, they may be nonnegligible up to a few wavelengths outside the dielectric.

A low-permittivity, low-flare-angle dielectric cone, excited in the $HE_{1,1}$ mode by a relatively inexpensive launching horn such as a metallic conical horn, can provide a highly symmetric pattern, with a sidelobe level of about 16 to 18 dB below boresight. With a more sophisticated (and expensive) launcher such as a corrugated horn the sidelobe level is significantly reduced. Alternatively, the use of a metal shield around the dielectric cone provides a simple means of reducing the sidelobes to a comparable low level, about 25 dB below boresight or less, even with a simple launcher. With proper design, the effect of the frequency on the radiation pattern of a low-permittivity dielectric cone can be minimized over a fairly large bandwidth.

Even if the dominant $HE_{1,1}$ is the most useful mode for antenna work, other modes may affect the aperture fields, the radiation pattern, and the cross-polarization. It is possible to minimize the influence of those modes by judiciously choosing the launcher or simply by using a dielectric of low enough permittivity to ensure that the higher-order modes are below cutoff.

The modal theory cannot explain the relation between the maximum cone semiflare angle and the dielectric permittivity. The basic relation between these two design parameters was derived from simple geometric optics, on the line of the "trapping region" of the dielectric cylinder.

With usual values of permittivity, dielectric cones with medium to large semiflare angles become a practical proposition. Adding a metallic shield the resulting performance is comparable to that of a corrugated horn of the same length and semiflare angle at a fraction of the cost.

The dielectric cone may be used to provide fairly large values of directivity, about 23 to 25 dBi, with low sidelobe levels and low cross-polarization. However, this requires a large aperture and a low phase error that, like any other horn, imply a low horn flare angle and thus a long, possibly impractical, horn. In addition, a long and heavy launcher is likely to be required. To reduce the cone and launcher size we may increase the dielectric cone flare angle but then we end up with a large aperture phase error and a correspondingly lower directivity. Dielectric lenses (the subject of Chapter 7) on the cone aperture may be used to correct the aperture phase error and provide a much less bulkier alternative to the low-flare-angle cone.

REFERENCES

[1] Bartlett, H. E., and Moseley, R. E., (1966), Dielguides — Highly Efficient Low Noise Antenna Feeds, *Microwave Journal*, Vol. 9, pp. 53–58.

[2] Clarricoats, P. J. B., (1970), Similarities in the Electromagnetic Behaviour of Optical Waveguides and Corrugated Feeds, *Electronics Letters*, Vol. 6, No. 6, pp. 178–180.

[3] Clarricoats, P. J. B., and Salema, C. E. R. C., (1971), Propagation and Radiation Characteristics of Low Permittivity Dielectric Cones, *Electronics Letters*, Vol. 7, No. 17, pp. 483–485.

[4] Clarricoats, P. J. B., and Salema, C. E. R. C., (1972), Influence of the Launching Horn on the Radiation Characteristics of a Dielectric Cone Feed, *Electronics Letters*, Vol. 8, No. 8, pp. 200–202.

[5] Salema, C. E. R. C., and Clarricoats, P. J. B., (1972), Radiation Characteristics of Dielectric Cones, *Electronics Letters*, Vol. 8, No. 16.

[6] Salema, C. E. R. C., (1972), *Theory and Design of Dielectric Cone Antennas*, Ph.D. Thesis, University of London.

[7] Clarricoats, P. J. B., and Salema, C. E. R. C., (1973) Antennas Employing Conical Dielectric Horns Part I, *Proceedings of the IEE*, Vol. 120, pp. 741–749.

[8] Clarricoats, P. J. B. and Salema, C. E. R. C, (1973), Antennas Employing Conical Dielectric Horns Part II, *Proceedings of the IEE*, Vol. 120, pp. 750–756.

[9] Lier, E., (1986), A Dielectric Hybrid Mode Antenna Feed: A Simple Alternative to the Corrugated Horn, *IEEE Transactions on Antennas and Propagation*, Vol. AP-34, No. 1, pp. 21–29.

[10] Olver, A. D., Clarricoats, P. J. B., Kishk, A. A., and Shafai, L., (1994), *Microwave Horns and Feeds*, IEEE Press, New York.

[11] Chatterjee, R., (1985), *Dielectric and Dielectric Loaded Antennas*, Research Studies Press, Letchworth, Hertfordshire, England.

[12] Harrington, R. F., (1961), *Time Harmonic Electromagnetic Fields*, McGraw-Hill, New York.

[13] Angot, A., (1961), *Complements de Mathematiques à l'usage des ingénieurs de l'electrotechnique et des télécommunications*, Editions de la Revue d'Optique, Paris.

[14] Saha, P. K., (1970), *Propagation and Radiation Characteristics of Corrugated Waveguides*, Ph.D. Thesis, University of Leeds.

[15] Abramowitz, M., and Stegun, I., (ed.), (1965), *Handbook of Mathematical Functions*, Dover Publications, New York.

Chapter 7

Dielectric Lenses

7.1 INTRODUCTION

Dielectric lenses may be used in combination with waveguides or horns to modify their radiation characteristics.

Due to the lower propagation velocity, phase error at the plane aperture of a solid dielectric horn is higher than for a conventional metallic horn of comparable size, thus placing a lower limit to the maximum achievable gain. Following a standard approach for high-performance metallic horns, where a dielectric lens is positioned at the horn aperture to correct the phase error [1, 2, 3], the dielectric horn aperture may be shaped into a lens [1].

There are other applications where the issue is not antenna gain, but rather beam shape. Again a lens may be used to modify the basic dielectric horn radiation pattern.

Intrinsic to the dielectric horn-lens geometry, there is only one refracting surface and so amplitude shaping and phase correction conditions can not be imposed simultaneously. The situation is different for metallic horns where correction lenses have two refracting surfaces, one on the horn side and the other on the output side, enabling two independent conditions to be met. In this case, however, rather than designing the second refracting surface for an amplitude condition it is often better to set the input surface as planar and to design the output surface according to the path length condition, since an appropriate choice of permittivity leads to convenient field distributions on the aperture [1].

Another strategy could be to employ moderate loss materials to reduce field amplitude toward the lens axis where its depth increases, thus reducing the field taper across the aperture without interfering with the other lens surface [4] at the expense of gain.

Horn lenses are just one possible application of dielectric lenses,a subject which is rather well covered in a number of antenna books [5, 6, 7, 8, 9, 10, 11]. Most cases deal with axis-symmetric lenses, designed as collimating devices, possibly with more than one focal point for scanning and multibeam applications. In the simplest designs for single focus lenses one of the two lens surfaces is arbitrarily fixed to some preferred shape while phase correction condition defines the other surface. For scanning and multibeam applications the second surface is used to introduce a further condition to minimize aberrations originated by off-axis displacements of the feed position.

Far less work is reported on dielectric lens design incorporating amplitude shaping conditions. One motivation is to produce an adequate field distribution over the lens aperture to improve the aperture efficiency and to reduce the sidelobe level [12, 13, 14, 15, 16, 17]. Routines to calculate two surface lenses for simultaneous phase and amplitude shaping are given in [10]. Another motivation is to shape the output beam into an hemispherical or a secant square ($\sec^2$) pattern for constant flux applications. A renewed interest on these type of patterns comes from emerging millimeter waves mobile broadband cellular systems and wireless local area network applications, where even nonsymmetric lenses may be required [18, 19, 20, 21].

It seems that the only paper dealing analytically with arbitrary 3D dielectric lens surface shaping is due to Westcott [16] who used complex coordinates to obtain an elegant and flexible formulation that requires no symmetry assumptions. An approximate method to design a nonsymmetric lens for a mobile communications application, together with theoretical and experimental results, is presented in [19].

In this chapter we deal with single-surface homogeneous dielectric lenses shaped on the aperture of solid dielectric horns. The analysis is extended up to a horn semiflare angle of 90 degrees. In this case the horn no longer exists and the lens modifies the radiation pattern of the waveguide feed. Except for this rather extreme case, the solid dielectric horn is enclosed in a metallic shield as discussed in Chapters 4 and 6 to ensure that radiation comes only from the front aperture.

Lens design, based on Geometrical Optics, is addressed either for aperture phase error correction or for output beam shaping. For phase correction the formulation is restricted to circular symmetric geometries, but results are adapted to suit pyramidal dielectric horns as well. A general formulation is presented for amplitude shaping that allows for nonrotational symmetric lenses.

7.2 PHASE CORRECTION

7.2.1 Lens Design

Although modal analysis used in previous chapters provides a satisfactory description of solid dielectric horn aperture fields, it is more convenient to turn to geometric optics for shaping the horn aperture into a lens.

Consider the geometry of Figure 7.1, which represents an axis-symmetric homogeneous solid dielectric horn with shaped aperture. Starting from the spherical wave front inside the dielectric horn we define associated rays originating at point O and refracting at surface $r(\theta)$ according to the Snell's laws.

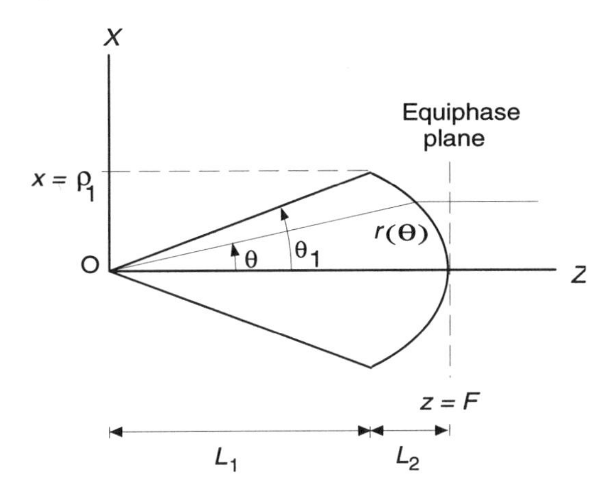

Figure 7.1: Geometry of axis-symmetric solid dielectric horn with shaped aperture for phase correction.

The optical path along the rays from the origin O to plane $z = F$ must be constant for all θ and φ:

$$r(\theta)\bar{\beta} + F - r(\theta)\cos\theta = F\bar{\beta} \tag{7.1}$$

where $\bar{\beta} = \beta/k_0$ is the normalized longitudinal propagation constant.

Equation (7.1) is written as if $\bar{\beta}$ were constant with z. Actually $\bar{\beta}$ changes with z but this affects all the rays in the same way. The differential phase between the rays results only from the lens shape, in a region where the horn cross section dimensions are expected to be much larger than λ where $\bar{\beta}$ is almost constant and approaches $\sqrt{\varepsilon_1}$. Under this assumption, rearranging (7.1) we obtain

$$r(\theta) = \frac{F(\sqrt{\varepsilon_1} - 1)}{\sqrt{\varepsilon_1} - \cos\theta} \tag{7.2}$$

Equation (7.2) defines an ellipse with major axis coaxial with the Z axis, farther focus coincident with the horn apex, and minor to major axis ratio given by

$$\frac{b}{a} = \sqrt{1 - \frac{1}{\varepsilon_1}} \tag{7.3}$$

It may be helpful to express F in terms of the horn semi-flare angle θ_1 and the aperture radius ρ_1:

$$F = \frac{\rho_1}{\sin\theta_1}\left[\frac{\sqrt{\bar{\varepsilon}_1} - \cos\theta_1}{\sqrt{\bar{\varepsilon}_1} - 1}\right] \tag{7.4}$$

Figure 7.2 shows the lens depth and lens-horn depth to aperture diameter ratio $L_2/2\rho_1$ and $F/2\rho_1$ versus horn semiflare angle θ_1. For the same flare angle, $L_2/2\rho_1$ and $F/2\rho_1$ are larger for smaller values of $\bar{\varepsilon_1}$. Also for smaller values of $\bar{\varepsilon_1}$ the ratio $L_2/2\rho_1$ grows faster with θ_1. The ratio $F/2\rho_1$ becomes more favorable as θ_1 increases, but there is an upper limit for this angle for two sorts of reasons. For one, θ_1 must not exceed the value that allows the propagation of higher-order modes in the dielectric horn. According to Chapter 4 this could be of the order of $\theta_1 < 12^o$ for $\bar{\varepsilon_1} = 2.53$ and $\theta_1 < 54.1^o$ for $\bar{\varepsilon_1} = 1.1$.

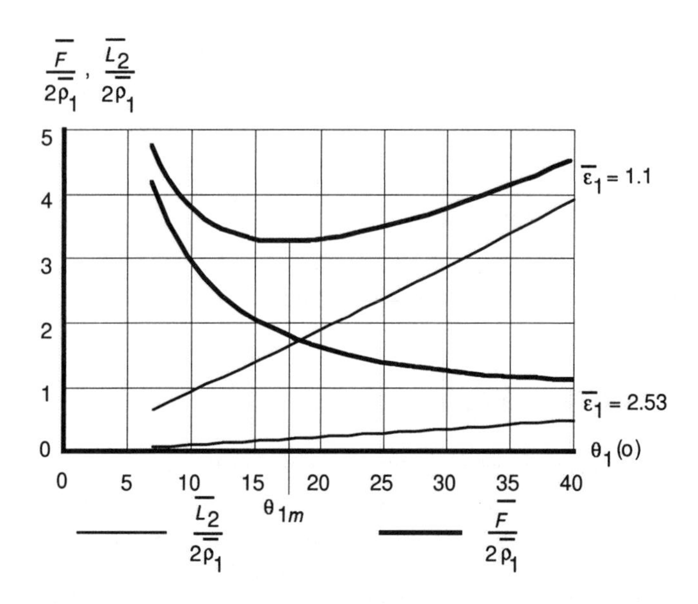

Figure 7.2: Lens depth and lens-horn depth to aperture ratio versus horn semiflare angle θ_1, for $\bar{\varepsilon}_1 = 1.1$ and $\bar{\varepsilon_1} = 2.53$.

The second reason is that the lens depth shall not exceed the ellipse major semiaxis $L_2 \leq F\sqrt{\bar{\varepsilon}_1}/(\sqrt{\bar{\varepsilon}_1} + 1)$ that is

$$\theta_{1m} \leq \arccos\left(\frac{1}{\sqrt{\bar{\varepsilon}_1}}\right) \tag{7.5}$$

which happens to coincide with the limit established in Section 6.5 for the maximum cone semiflare angle.

For $\bar{\varepsilon}_1 = 2.53$ we obtain $\theta_{1m} = 51$ degrees, far beyond the single mode operation condition. For $\bar{\varepsilon}_1 = 1.1$ we get from (7.5) $\theta_{1m} = 17.5$ degrees a limit lower than the multimode threshold.

7.2.2 Application to the Solid Dielectric Pyramidal Horn

The results from the previous section are not generally applicable to the pyramidal horn, due to its lack of circular symmetry and noncoincident E- and H-plane apexes. However, they may be used for pyramidal horns with approximately square cross sections (and similar E- and H-plane flare angles) that produce equal main lobe radiation patterns in both planes.

Figure 7.3 represents a lens of revolution generated by the profile given by (7.2) machined in the cross section of a solid dielectric pyramidal horn with $\bar{\varepsilon}_1 = 2.53$. The lens diameter ρ_1 equals the diagonal of the horn square cross section with normalized dimensions $\bar{a}_p = \bar{b}_p = 40$ and flare angles $\alpha_E = 20.2$ degrees and $\alpha_H = 18.8$ degrees. The focus of the lens coincides with the horn apex in the H-plane. The dielectric lens-horn is inserted into a pyramidal metallic shield with aperture dimensions $\bar{a}_m = \bar{b}_m = 47.8$.

This arrangement should radiate roughly as a plane aperture with the same dimensions, a cosine–cosine illumination and zero phase error which, from Section 2.11.2, yields a sidelobe level of −23 dB (below boresight), a directivity of 25.2 dBi, and an aperture efficiency of 65.7%.

Figure 7.3: Lens with revolution symmetry shaped in the aperture of a solid dielectric horn of square cross section.

Figure 7.4 shows the computed and measured radiation patterns with lens correction at 62.5 GHz. As in previous chapters the computed patterns are obtained from (2.131) where the integration surface coincides with the lens and extends to the metal shield walls. Field amplitudes are given in Section 3.2.2 with propagation constants defined in Section 3.2.4

for the mTE to X mode, after taking refraction at the lens aperture into account. Phase is calculated in line with the optical rays principles.

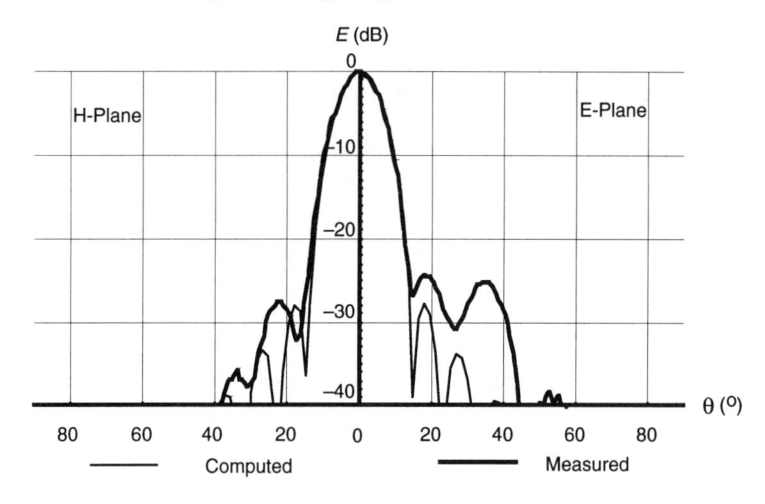

Figure 7.4: (1) Computed and (2) measured patterns of a lens-corrected shielded solid dielectric pyramidal horn with $\bar{\varepsilon_1} = 2.53$, $\bar{a_p} = \bar{b_p} = 40$, $\alpha_E = 20.2$ degrees, and $\alpha_H = 18.9$ degrees, fed by standard V-band metallic waveguide ($\bar{a} = 4.84$ and $\bar{b} = 2.48$).

Measured values of sidelobe level, directivity, and aperture efficiency are −24.3 dB below boresight, 24.6 dBi, and 56.8%, respectively, closely matching the values predicted from the plane aperture with a cosine–cosine illumination and zero phase error. The pattern symmetry is good in the main lobe but deteriorates in the sidelobe region due to the presence of unwanted higher-order modes.

Since the lens is designed for the fundamental mode it defocuses higher-order modes due to their different propagation velocity, and thus it decreases their importance in the main lobe.

The dielectric losses, and the internal reflection at the lens aperture contribute to reduce gain. An upper estimate of material attenuation in dB can be obtained from

$$A_m = 4.3\sqrt{\bar{\varepsilon}_1}\tan(\delta)\,k_0F \tag{7.6}$$

For commercially available materials with $\bar{\varepsilon}_1 = 2.53$ and $\tan(\delta) = 0.0005$ we obtain $A_m = 0.44$ dB for the example above, where $k_0F = 128.86$. Reflection losses at the lens surface can be reduced by using standard (narrow band) $\lambda/4$ matching layer techniques.

Near the apex the solid dielectric horn is shaped into a dielectric waveguide that fits inside the metallic waveguide feeder and ends in the form of a wedge to favor impedance matching at the waveguide port.

As shown in Figure 7.5 measured reflection loss is below -17 dB for a moderately large bandwidth even if the lens increases the input reflections, due to the focusing effect on the backward-reflected rays.

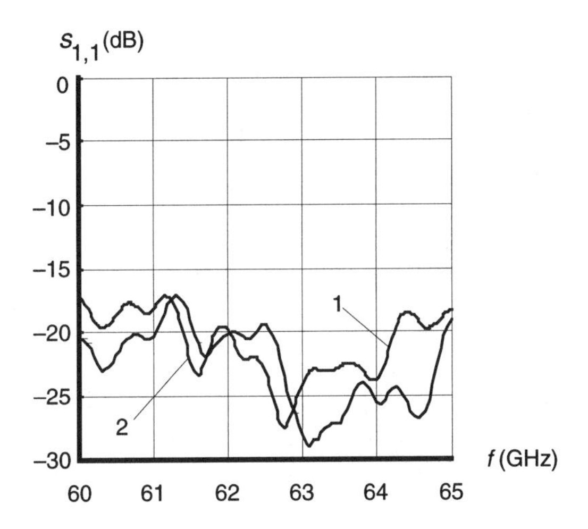

Figure 7.5: Reflection loss of the shielded solid dielectric horn measured at the waveguide port (1) with and (2) without lens.

7.2.3 Application to the Solid Dielectric Conical Horn

In Figure 7.6 we represent the radiation patterns of a lens-corrected solid dielectric conical horn with $\bar{\rho}_1 = 17.6$ and $\theta_1 = 18$ degrees inserted into a metallic conical shield with $\bar{\rho}_{1m} = 27.8$ and $\theta_{1m} = 25$ degrees. Measured directivity is 22.5 dBi,which compares with 23.5 dBi given by (2.195) for a circular aperture with a cosine illumination and no phase error. Ripple, imperfect circular symmetry, and somewhat higher than expected sidelobe levels on the measured patterns are due to higher-order modes.

7.3 AMPLITUDE SHAPING

7.3.1 Lens Design

The objective is to find the three-dimensional shape of a solid dielectric horn aperture that yields a prescribed output radiation pattern G given the feed power pattern U inside the dielectric horn. We assume that the aperture dimensions and the principal radii of

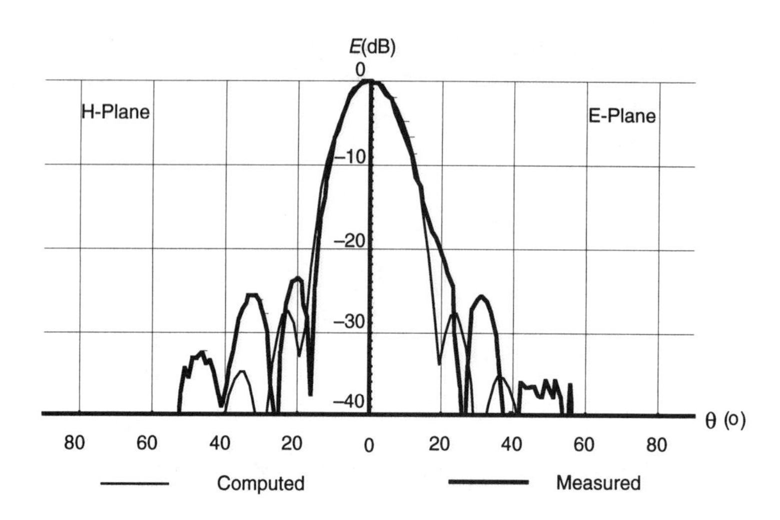

Figure 7.6: (2) Measured and (1) computed radiation patterns for lens corrected conical solid dielectric horn with $\bar{\varepsilon}_1 = 2.53$, $\bar{\rho}_1 = 17.6$, and $\theta_1 = 18$ degrees at 62.5 GHz.

curvature at every point on the aperture are much larger than the wavelength λ so that geometric optics may be used for the design.

Consider the geometry of Figure 7.7. An elementary ray tube with its origin at the horn apex O is associated with the direction $\boldsymbol{i}$ (angles θ, ϕ). The intersection of this ray tube with the horn 3D aperture $r(\theta, \phi)$ defines an elementary surface dS with unit normal $\boldsymbol{n}$. The power associated with the incident wave in this ray tube is reflected and transmitted at the interface. The associated directions are $\boldsymbol{r}$ and $\boldsymbol{t}$, respectively.

An appropriate angle transformation $\theta'(\theta, \phi)$, $\phi'(\theta, \phi)$ is to be found by imposing power conservation in the elementary ray tube. When applying Snell's law this transformation must yield a lens surface that is physically viable.

Power conservation along the elementary ray tube yields

$$U(\theta, \phi)T \sin\theta \, d\theta \, d\phi = KG(\theta', \phi') \sin\theta' \, d\theta' \, d\phi' \tag{7.7}$$

Function T represents the ratio of the transmitted power P_t across dS to the incident power P_i, and is defined in (2.115). K is a normalization constant to be determined from the balance between the total power inside and outside the horn:

$$K = \frac{\int_0^{2\pi} \int_0^{\theta_M} U(\theta, \phi)T \sin\theta \, d\theta \, d\phi}{\int_0^{2\pi} \int_0^{\theta'_M} G(\theta', \phi') \sin\theta' \, d\theta' \, d\phi'} \tag{7.8}$$

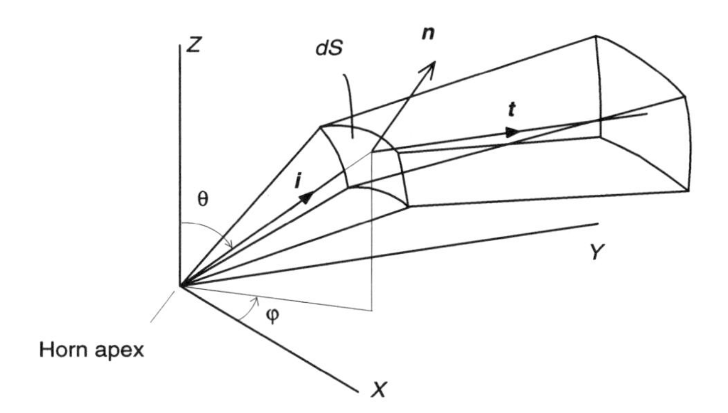

Figure 7.7: Geometry of the ray tube associated with incidence and refraction directions.

Using the Jacobian of transformation $\theta'(\theta, \phi)$, $\phi'(\theta, \phi)$, (7.7) may be written as

$$U(\theta, \phi)T \sin\theta = KG(\theta', \phi') \sin(\theta') \left(\frac{\partial\theta'}{\partial\theta}\frac{\partial\phi'}{\partial\phi} - \frac{\partial\theta'}{\partial\phi}\frac{\partial\phi'}{\partial\theta} \right) \tag{7.9}$$

The transformation $\theta'(\theta, \phi)$, $\phi'(\theta, \phi)$ defines the transmitted ray direction $\boldsymbol{t}$ required for each incident ray direction $\boldsymbol{i}$. Application of Snell's laws yields the surface slopes that agree with pairs $(\boldsymbol{i}, \boldsymbol{t})$. Snell's refraction law may be written as

$$\left(\sqrt{\varepsilon_1}\boldsymbol{i} - \boldsymbol{t}\right) \times \boldsymbol{N} = 0 \tag{7.10}$$

where $\boldsymbol{N}$ is the surface normal at the incidence point

$$\boldsymbol{N} = \boldsymbol{i} - \frac{1}{r}\frac{\partial r}{\partial\theta}\boldsymbol{\theta} - \frac{1}{r\sin\theta}\frac{\partial r}{\partial\phi}\boldsymbol{\phi} \tag{7.11}$$

Expanding (7.10) and using (7.11) leads to the following system of equations:

$$\frac{\partial r}{\partial\theta} = \frac{t_\theta r}{\sqrt{\varepsilon_1} - t_r} \tag{7.12}$$

$$\frac{\partial r}{\partial\phi} = \frac{t_\phi r \sin\theta}{\sqrt{\varepsilon_1} - t_r} \tag{7.13}$$

where (t_r, t_θ, t_ϕ) represent the components of unit vector $\boldsymbol{t}$ expressed in the coordinate frame of the feed:

$$t_r = \sin(\theta')\sin(\theta)\cos(\phi' - \phi) + \cos(\theta')\cos(\theta) \quad (7.14)$$

$$t_\theta = \sin(\theta')\cos(\theta)\cos(\phi' - \phi) - \cos(\theta')\sin(\theta) \quad (7.15)$$

$$t_\phi = \sin(\theta')\sin(\phi' - \phi) \quad (7.16)$$

A further condition ensures a smooth surface:

$$\frac{\partial}{\partial\theta}\left(\frac{\partial r}{\partial\phi}\right) = \frac{\partial}{\partial\phi}\left(\frac{\partial r}{\partial\theta}\right) \quad (7.17)$$

Substituting (7.12) and (7.13) into (7.17) and rearranging gives

$$a_c\frac{\partial\theta'}{\partial\phi} + b_c\frac{\partial\phi'}{\partial\phi} + c_c\frac{\partial\theta'}{\partial\theta} + d_c\frac{\partial\phi'}{\partial\theta} = 0 \quad (7.18)$$

where

$$a_c = \cos(\phi' - \phi)\left[\sqrt{\bar{\varepsilon}_1}\cos(\theta)\cos(\theta') - 1\right] + \sqrt{\bar{\varepsilon}_1}\sin(\theta)\sin(\theta') \quad (7.19)$$

$$b_c = -\sin(\theta')\sin(\phi' - \phi)\left[\sqrt{\bar{\varepsilon}_1}\cos(\theta) - \cos(\theta')\right] \quad (7.20)$$

$$c_c = -\sin(\theta)\sin(\phi' - \phi)\left[\sqrt{\bar{\varepsilon}_1}\cos(\theta') - \cos(\theta)\right] \quad (7.21)$$

$$d_c = -\sin(\theta)\sin(\theta')\left[\left(\sqrt{\bar{\varepsilon}_1} - \cos(\theta')\cos(\theta)\right)\cos(\phi' - \phi) - \sin(\theta)\sin(\theta')\right] \quad (7.22)$$

Finding the surface profile $r(\theta, \phi)$ involves solving the system of partial differential equations formed by (7.9), (7.12), (7.13), and (7.18). These equations are formally of the same type as those described in [22, 23] for the synthesis of reflector antennas so, in principle they should be solved in a manner similar to the one put forward in [22].

Alternatively it should be possible to formulate this problem as a second-order nonlinear partial differential equation of the Monge-Ampere type. This has been done in [16] using the method of complex coordinates, although this is not strictly necessary to obtain the Monge-Ampere equation [22]. The numerical solution of this type of equations is discussed in [24].

Whatever the formulation used and the accuracy of the calculated $r(\theta, \phi)$ some discrepancy should be expected between the resulting radiation pattern and the target pattern. Geometrical optics assumptions may be violated, at least in some parts of the aperture, where the size of the principal radii of curvature of the lens is comparable with λ. On the other side diffraction was not considered. Some optimization of the aperture shape may be required in a second design step [25].

7.3.2 Axis-Symmetric Case

The formulation presented in the previous section may be simplified when $\partial/\partial\phi = 0$, that is, when both the waveguide feed pattern and the target pattern are axis-symmetric. In this case $\phi' = \phi$ and, consequently

$$\begin{aligned}\partial\phi'/\partial\phi &= 1\\ \partial\theta'/\partial\phi &= 0\\ \partial r/\partial\phi &= 0\end{aligned}$$

The design equations (7.9), (7.12), and (7.13) reduce to

$$\begin{aligned}\frac{d\theta'}{d\theta} &= \frac{T}{K}\frac{U(\theta)}{G(\theta')}\frac{\sin\theta}{\sin\theta'}\\ \frac{dr}{d\theta} &= \frac{r\sin(\theta'-\theta)}{\sqrt{\bar{\varepsilon}_1}-\cos(\theta'-\theta)}\end{aligned} \tag{7.23}$$

In the examples that follow differential equations (7.23) are solved using the second-order Runge-Kutta formulas. The initial values taken for angles θ and θ' correspond respectively to the horn semiflare angle θ_1 and to half the output beamwidth .

The normalization factor K in (7.23) depends indirectly on the unknown function $r(\theta)$ through an integral relation (7.8). If medium losses are neglected and transmittance T at the lens output surface is assumed constant for all values of θ, then K becomes independent of the lens shape allowing its calculation prior to determining the lens profile. The solution for $r(\theta)$ can then be obtained from (7.23) integrating progressively from the lens aperture edge to its apex or alternatively from the center to the edge. Calculations are very fast and do not require iterations.

If dielectric losses and reflections at the lens surface are to be considered then an iterative procedure follows the previous step. Taking the calculated profile $r(\theta)$ we evaluate $T(\theta)$ and the power loss inside the lens as a function of θ. With these results we compute a new solution of (7.23). The new profile $r(\theta)$ may be used for another iteration if necessary although two steps are usually enough.

Figure 7.8 shows the shaped aperture of a shielded dielectric conical horn calculated with the design target to produce a flat top radiation pattern for $\theta'_{max} < 33$ degrees. The flare angle is $\theta_1 = 17$ degrees and the aperture diameter is of the order of 11.5λ. The launcher is a circular metallic waveguide operated in the $TE_{1,1}$ mode with circular polarization.

Figure 7.9 shows the corresponding radiation pattern, calculated for $f = 62.5$ GHz. A reasonably constant radiation pattern is obtained for $\theta < 26$ degrees. The amplitude ripple of about ± 0.5 dB is due to weak diffraction at aperture edge.

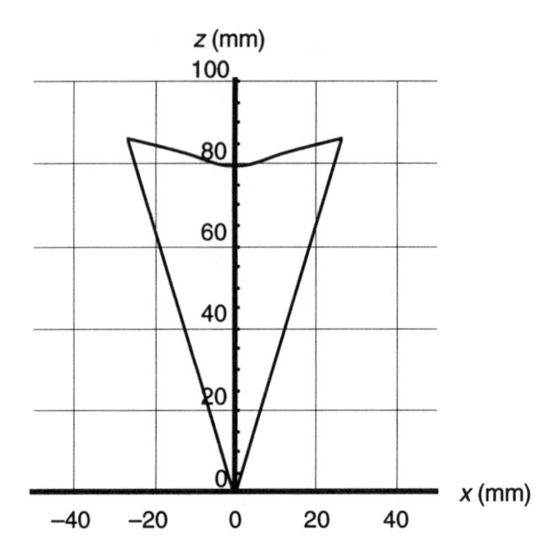

Figure 7.8: Profile of a lens-corrected shielded conical dielectric horn with $\theta_1 = 17$ degrees designed for flat top output beam.

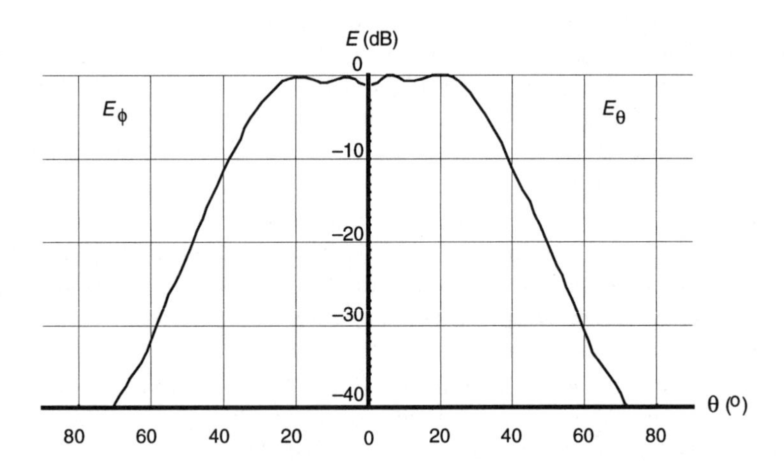

Figure 7.9: Calculated radiation pattern at $f = 62.5$ GHz of a lens-corrected shielded conical dielectric horn with $\theta_1 = 17$ degrees designed for flat top output beam.

For constant horn flare angle θ the radiation pattern decay past the flat region is steeper for θ'_{max} angles closer to θ_1, and for increasing aperture sizes.

Implicit shortcomings of this configuration are:

- The value of θ'_{max} is indirectly limited by the maximum value of θ_1, which avoids exciting higher-order modes;
- A large aperture implies a corresponding increase in horn length, to comply with the restrictions on θ_1.

A radical solution for these shortcomings is presented in the following section for these cases.

7.3.3 Wide-Angle Radiation Pattern

There are applications where a single beam is required to extend to $\theta'_{max} \approx 75$ degrees and beyond. For instance in some wireless millimeter wave communications systems the mobile terminal antenna should have an almost hemispherical pattern and the base station antenna a cosecant square ($\csc^2$) pattern over a very wide angular interval to ensure constant received power in the coverage area.

Given the constrains referred in the previous section, the alternative is to increase the horn flare angle θ_1 beyond the limit where there is no longer a guided wave. In the examples that follow we took $\theta_1 = 90$ degrees and used no metallic shield, so the waveguide aperture radiates as in an unbounded dielectric media.

This can be confirmed in Figure 7.10 where we compare computed and measured radiation patterns measurements of an open-ended metallic waveguide, with radius $\rho_g = 1.85$ mm and wall thickness of 0.5 mm operated in the $\text{TE}_{1,1}$ mode with circular polarization at 62.5 GHz, radiating into a dielectric medium with $\bar{\varepsilon}_1 = 2.53$. Computed patterns were obtained from from standard waveguide aperture radiation formulas.

Since the radiation pattern inside a solid dielectric cannot be measured directly we terminated the waveguide with an hemispherical dielectric lens of radius $R = 6\lambda$ and measured the tangential field components outside the dielectric at a radial distance $d \approx 1.5\lambda$ away from the surface. The continuity of the tangential fields components across the dielectric-air interface and the fact that this surface is everywhere normal to the direction of propagation of the wave radiated from the waveguide aperture ensure that the measured fields are an accurate description of the open-ended waveguide radiation pattern inside the dielectric.

Radiation from the waveguide aperture into an unbounded media is a canonical problem that has received considerable attention (see [5], Chapter 7 of [1], and [26, 27, 28, 29, 30, 31]). However, to demonstrate the point it seems enough to consider the Risser formulas [5] derived from aperture integration using the Stratton-Chu formulation adapted

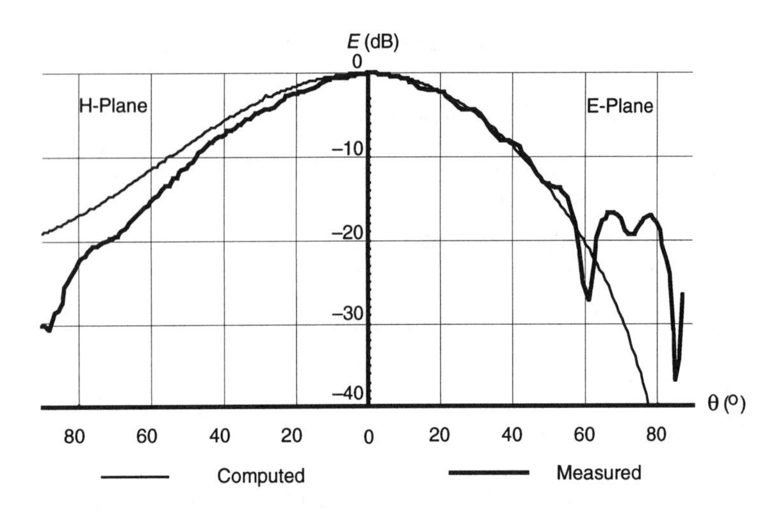

Figure 7.10: (1) Computed and (2) measured radiation pattern of an open-ended circular waveguide operated in the $TE_{1,1}$ mode, with circular polarization, at 62.5 GHz, radiating into a dielectric medium with $\bar{\varepsilon}_1 = 2.53$.

for $\bar{\varepsilon_1} \neq 1$.The agreement between measured and computed results in Figure 7.10 is remarkable for the E_θ component up to 20 dB below the maximum, while for the E_ϕ component equations from [5] overestimate the result. Similar conclusions were obtained using the same procedure for other waveguide modes and feeds.

An interesting result is the very good symmetry between measured E_θ and E_ϕ components. This is a necessary condition to obtain satisfactory symmetry in the output pattern of the lens excited by this feed.

For comparison, Figure 7.11 shows the half-lens profile and Figure 7.12 the radiation pattern for the same pattern specification as in the example of the previous section. The feed is the same referred to in Figure 7.12.

For comparable maximum aperture diameter, the lens depth is much less than the horn considered in the example of the previous section. The ripple on the flat region is slightly higher than before because the field amplitude at the aperture edge is higher. This can be improved by very slightly increasing the waveguide aperture radius. The decay ratio of the field past the flat region is steeper for the lens. This may be further improved by scaling up the lens.

The most important advantage of this structure is for wide-angle applications of which two examples are given in Figures 7.13 and 7.14 for a flat top and a cosecant square pattern, respectively, where the desired patterns are extended up to $\theta' = 76$ degrees. Again the lens is excited by the $TE_{1,1}$ mode with circular polarization, but the waveguide radius is reduced

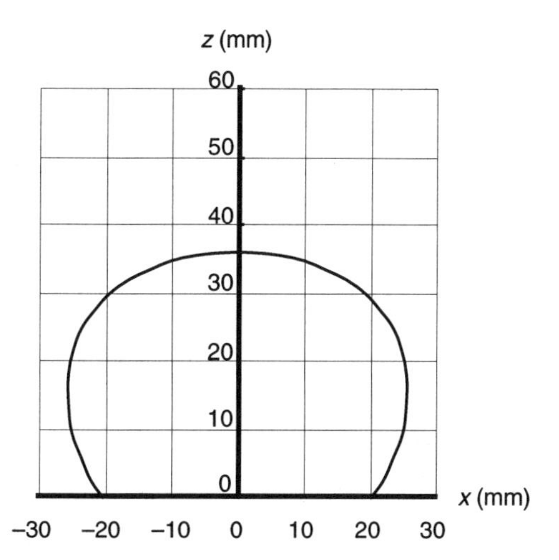

Figure 7.11: Profile of a dielectric lens designed to produce a flat top output beam from a metallic waveguide operated in the $TE_{1,1}$ mode with circular polarization.

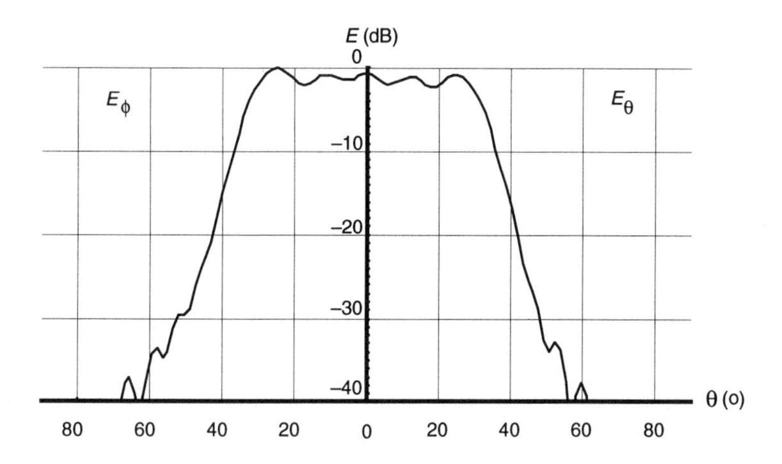

Figure 7.12: Calculated radiation pattern of a dielectric lens designed to produce a flat top output beam from a metallic waveguide operated in the $TE_{1,1}$ mode with circular polarization at $f = 62.5$ GHz.

to $\rho_g = 1.5$ mm to improve the lens aperture illumination. Ripple in the calculated and measured patterns is both due to the spillover from the waveguide and to multiple internal reflections not accounted for in the calculated patterns.

Matching layers may not be effective for these lenses since in much of the lenses' output surface the rays emerge at widely varying angles, often far away from the normal. A procedure to minimize the adverse effects of reflections is described in [21].

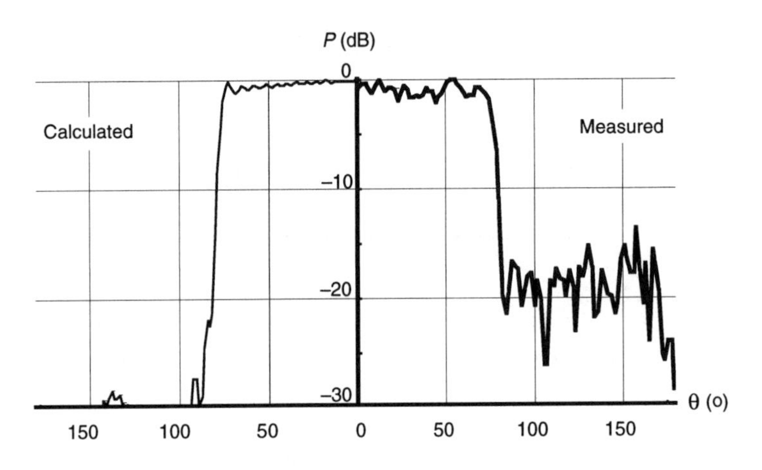

Figure 7.13: Calculated and measured power patterns of a 66-mm-diameter dielectric lens with $\bar{\varepsilon}_1 = 2.53$ designed for a constant power pattern within $\theta < 76$ degrees and fed by the $TE_{1,1}$ mode of a circular waveguide with circular polarization, at $f = 62.5$ GHz.

Unlike previous cases, the 135-mm-diameter lens designed for the cosecant square pattern in Figure 7.14, which was obtained from the formulation of Section 7.3.2, must be corrected to account for stronger diffraction effects. Figure 7.15 shows the profile obtained from geometrical optics before and after corrections. The corresponding calculated patterns are plotted in Figure 7.16.

The two lens profiles are coincident for high values of θ but gradually separate as θ decreases to zero. It can be shown that where both profiles are coincident they approach an ellipse with major axis directed along θ'_{max} and farther focus located at the feed point. In fact both patterns in Figure 7.16 comply with the θ'_{max} condition. Correction in the valley region of the lens to divert the end-fire radiation of the feed leads the refracted rays to emerge almost tangent to the lens surface. The depression of the corrected profile forces more energy from the feed to be refracted away from the end-fire direction [20].

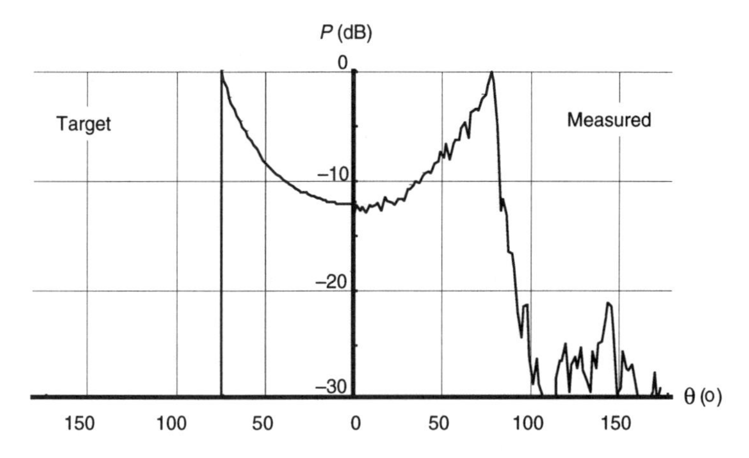

Figure 7.14: Power patterns of a 135-mm-diameter dielectric lens with $\bar{\varepsilon}_1 = 2.53$ designed for a cosecant square pattern within $\theta = 76$ degrees and fed by the $TE_{1,1}$ mode of a circular waveguide with circular polarization, at $f = 62.5$ GHz.

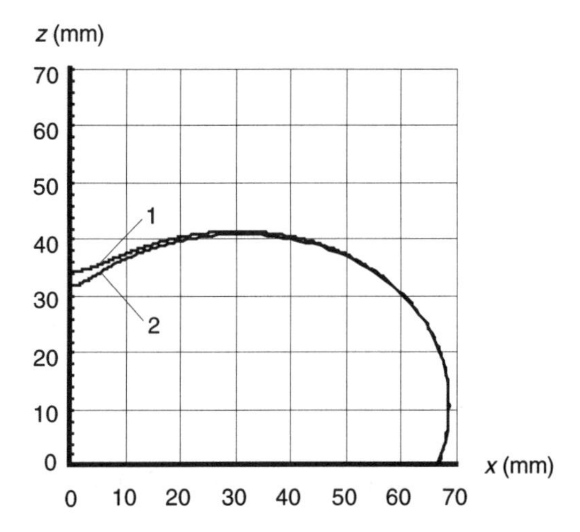

Figure 7.15: Profile of a dielectric lens designed for a cosecant square pattern using geometrical optics: (1) without correction for diffraction and (2) with correction for diffraction.

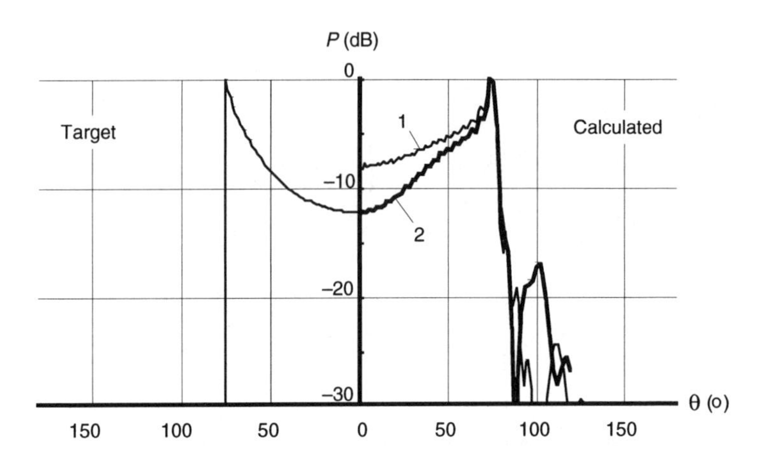

Figure 7.16: Calculated radiation power patterns using physical optics for a dielectric lens designed to produce a cosecant square pattern: (1) without correction for diffraction and (2) and with correction for diffraction.

7.4 CONCLUSIONS

The dielectric lens is a simple and effective way of modifying the radiation patterns of other antennas, particularly dielectric horns where the lens may be incorporated into the antenna. In this case since there is only a free surface — the output one — the goal is usually to correct aperture phase errors, thus increasing directivity. Although circular symmetry is often assumed in the analysis this does not prevent such lenses being used in conjunction with solid dielectric pyramidal horns with approximately square apertures. Phase correction lenses tend to reduce the contribution of higher-order modes to the main lobe.

Dielectric lenses may also be designed for amplitude shaping. In mobile communications a possible antenna design target is to provide a $\csc^2$ pattern over very wide and possibly dissimilar angles in the principal planes.

In practice,the conditions for single-mode operation restrict the lens diameter, which in turn limits the achievable amplitude shaping. Increasing the horn semiflare angle to 90 degrees eliminates this restriction.

Besides the $TE_{1,1}$ mode of a circular metallic waveguide with circular polarization, other types of waveguide modes and feeds may be used. For instance, the $TM_{0,1}$ circular waveguide mode and the TEM mode of the coaxial line have been employed where circular symmetric pattern with linear polarization are required [20, 21].

REFERENCES

[1] Olver A. D., Clarricoats, P. J. B., Kishk, A., and Shafai, L., (1994), *Microwave Horns and Feeds*, IEEE Press, New York.

[2] Clarricoats, P. J. B., and Saha, P. K., (1969), Radiation Patterns of a Lens-Corrected Conical Scalar Horn, *Electronic Letters*, Vol. 5, pp. 592–593.

[3] Kildal P. S., Jakobsen K., and Sudhakar R., (1984), Meniscus-Lens-Corrected Corrugated Horn: A Compact Feed for a Cassegrain Antenna, *IEE Proceedings*, Vol. 131, Part H, No. 6, pp. 390–394.

[4] Olver, A. D., Saleeb, A. A., (1979), Lens-type Compact Antenna Range, *Electronics Letters*, Vol. 15, No. 14, pp. 409–410.

[5] Silver, S., (ed.), (1949), *Microwave Antenna Theory and Design*, McGraw-Hill, New York.

[6] Brown, J., (1953), *Microwave Lenses*, Methuen, London.

[7] Fradin, A. Z., (1961), *Microwave antennas*, Pergamon Press, London.

[8] Jasik, H.,(1961), *Antenna Engineering Handbook*, McGraw-Hill, New York.

[9] Lo, Y. T. and Lee, S. W., (1988), *Antenna Handbook Theory Applications and Design*, Van Nostrand, New York.

[10] Sletten, C. J., (ed.), (1988), *Reflector and Lens Antennas - Analysis and Design Using Personal Computers*, Artech House, Norwood.

[11] Cornbleet, S., (1994), *Microwave and Geometrical Optics*, Academic Press, New York.

[12] Waineo, D., (1976), Lens Designed for Arbitrary Illumination, *IEEE Antennas and Propagation Symposium*, p. 476.

[13] Lee, J. J., (1982), A Dielectric Lens Shaped for a Generalized Taylor Distribution, *IEEE Antennas and Propagation Symposium*, p. 124.

[14] Lee, J. J., (1983), Dielectric Lens Shaping and Coma-Correction Zoning, Part I: Analysis, *IEEE Transactions on Antennas and Propagation*, Vol. AP-31, No.1, pp. 211–216.

[15] Lee, J. J., and Carlise, R. L., (1983), A Comma-Corrected Multibeam Shaped Lens Antenna, Part II: Experiments, *IEEE Transactions on Antennas and Propagation*, Vol. AP-31, No.1, pp. 216–220.

[16] Westcott, B. S., (1986), General Dielectric-Lens Shaping Using Complex Coordinates, *Proceedings of the IEE*, Vol. 133(H), pp. 122–126.

[17] Bishay, S., Cornbleet, S., and Hilton, J., (1989), Lens Antennas with Amplitude Shaping or Sine Condition, *Proceedings of the IEE*, Vol. 136(H), No. 3, pp. 276–279.

[18] Fernandes, C., Francês, P., and Barbosa, A., (1995), Shaped Coverage of Elongated Cells at Millimetrewaves Using Dielectric Lenses, *Proceedings of the 25th European Microwave Conference*, Bologna, Italy, pp. 66–70.

[19] Fernandes, C., Francês, P., and Barbosa, A., (1995), "Test Report on Antenna Assemblies," RACE 2067, Deliverable RACE Program - Project MBS, R2067/IST/4.6.2./DS/P/041.b1.

[20] Fernandes, C., Brankovic, V., Zimmermann, S., Filipe, M. and Anunciada, L., (1998), Dielectric Lens Antennas for Wireless Broadband Communications, to appear in *Wireless Personal Communications Magazine*, Special Issue on "Wireless Broadband Communications."

[21] Fernandes, C., Filipe, M., and Anunciada, L., (1997), Lens Antennas for the SAMBA Mobile Terminal, *Proceedings of the ACTS Mobile Telecommunications Summit 97*, Aalborg, Denmark, pp. 563–568.

[22] Galindo-Israel, V., Imbriale, W., and Mittra, R., (1987), On the Theory of the Synthesis of Single and Dual Offset Shaped Reflector Antennas, *IEEE Transactions on Antennas and Propagation*, Vol. AP-35, No. 8, pp. 887–896.

[23] Galindo-Israel, V., Imbriale, W., Mittra, R. and Shogen K., (1991), On the Theory of the Synthesis of Dual-Shaped Reflectors - Case Examples, *IEEE Transactions on Antennas and Propagation*, Vol. AP-39, No. 5, pp. 620–626.

[24] Westcott, B. S., (1993), *Shaped Reflector Antenna Design*, Research Studies Press, Letchworth, Hertfordshire, England.

[25] Lemaire, D., Fernandes, C., Barbosa, A., and Sobieski, P., (1996), A Method to Overcome the Limitations of the G.O. in Axis-Symmetric Dielectric Lens Shaping, *International Journal of Infrared and Millimetre Waves*, Vol. 17, No. 8, pp. 1377–1390.

[26] Teodoridis, V., Sphicopoulos, T., and Gardiol F., (1985), The Reflection from an Open-Ended Rectangular Waveguide Terminated by a Layered Dielectric Medium, *IEEE Transactions on Microwave Theory and Techniques*, Vol. MTT-33, No. 5, pp. 359–366.

[27] Gex-Fabry, M., Mosig, J., and Gardiol, F., (1979), Reflection and Radiation of an Open-Ended Circular Waveguide: Application to Nondestructive Measurement of Materials, *AE* , Vol. 33, No. 12, pp. 473–478.

[28] Lewin, L., (1951), *Advanced Theory of Waveguides*, Iliff, London

[29] Baudrand, H., Tao, J. W., and Atechian J., (1988), Study of Radiating Properties of Open-Ended Rectangular Waveguides, *IEEE Transactions on Antennas and Propagation*, Vol. AP-36, No. 8, pp. 1071–1077.

[30] Wu, D., and Kanda, M., (1989), Comparison of Theoretical and Experimental Data for the Near Field of an Open-Ended Rectangular Waveguide, *IEEE Transactions on Electrical Compatibility*, Vol. AP-31, No. 4, pp. 353–358.

[31] Moheb, H., and Shafai, L., (1990), Numerical Computation of Radiation Characteristics of Rectangular Waveguides, *Proceedings of the IEEE on Antennas and Propagation Symposium*, pp. 1843–1846.

Appendix A

Vector Analysis

A.1 GAUSS AND STOKES THEOREMS

The Gauss-Ostrogradsky theorem states that for any bounded volume V, enclosed by surface S, and any vector field $\mathcal{A}$ we have

$$\int_S (\mathcal{A} \cdot \boldsymbol{n})\, dS = \int_V (\nabla \cdot \mathcal{A})\, dV \tag{A.1}$$

where dS is the elementary area on S, dV the elementary volume in V, and $\boldsymbol{n}$ the unit vector outward normal to S.

Stokes's theorem states that in any bounded area S, enclosed by contour C, and any vector field $\mathcal{A}$ we have

$$\oint_C \mathcal{A} \cdot \mathrm{d}\boldsymbol{\ell} = \int_S \nabla \times \mathcal{A} \cdot \boldsymbol{n}\, dS \tag{A.2}$$

where $d\boldsymbol{\ell}$ is the elementary length along the contour, the positive sense being given by the so-called corkscrew rule — *positive direction is the same as a corkscrew has to turn in order to progress in the same sense as the normal* — and dS is the elementary area on S.

The element of length along the contour $d\boldsymbol{\ell}$ may also be written as

$$\mathrm{d}\boldsymbol{\ell} = \boldsymbol{\tau} d\ell \tag{A.3}$$

where $\boldsymbol{\tau}$ is the unit vector along the contour.

A.2 GENERAL IDENTITIES

Let us suppose that f is a scalar function and $\boldsymbol{A}$ and $\boldsymbol{B}$ are vector functions, each having continuous first and second derivatives everywhere within a volume V and on its boundary surface S. Then we have

$$
\begin{aligned}
\nabla \cdot (f\,\boldsymbol{A}) &= f\,\nabla \cdot \boldsymbol{A} + (\nabla f) \cdot \boldsymbol{A} && \text{(A.4)}\\
\nabla \times (f\boldsymbol{A}) &= f\nabla \times \boldsymbol{A} + (\nabla f) \times \boldsymbol{A} && \text{(A.5)}\\
\nabla \cdot (\boldsymbol{A} \times \boldsymbol{B}) &= \boldsymbol{B} \cdot (\nabla \times \boldsymbol{A}) - \boldsymbol{A} \cdot (\nabla \times \boldsymbol{B}) && \text{(A.6)}\\
\nabla \times (\boldsymbol{A} \times \boldsymbol{B}) &= (\nabla \cdot \boldsymbol{B})\boldsymbol{A} - (\nabla \cdot \boldsymbol{A})\boldsymbol{B} + (\boldsymbol{B} \cdot \nabla)\boldsymbol{A} - (\boldsymbol{A} \cdot \nabla)\boldsymbol{B} && \text{(A.7)}\\
\nabla(\boldsymbol{A} \cdot \boldsymbol{B}) &= \boldsymbol{A} \times (\nabla \times \boldsymbol{B}) + \boldsymbol{B} \times (\nabla \times \boldsymbol{A}) + (\boldsymbol{B} \cdot \nabla)\boldsymbol{A} + (\boldsymbol{A} \cdot \nabla)\boldsymbol{B} && \text{(A.8)}\\
\nabla \times (\nabla f) &= 0 && \text{(A.9)}\\
\nabla \cdot (\nabla \times \boldsymbol{A}) &= 0 && \text{(A.10)}\\
\nabla \times (\nabla \times \boldsymbol{A}) &= \nabla(\nabla \cdot \boldsymbol{A}) - \nabla^2 \boldsymbol{A} && \text{(A.11)}
\end{aligned}
$$

A.3 VECTOR OPERATORS AND THE POSITION VECTOR

Assume that $\boldsymbol{R}$ is the position vector of the general point P, with modulus R:

$$\boldsymbol{R} = R\boldsymbol{r} \qquad \text{(A.12)}$$

Then we have

$$
\begin{aligned}
\nabla R &= \boldsymbol{r} && \text{(A.13)}\\
\nabla \cdot \boldsymbol{R} &= 3 && \text{(A.14)}\\
\nabla \times \boldsymbol{R} &= 0 && \text{(A.15)}
\end{aligned}
$$

A.4 OPERATORS IN DIFFERENT COORDINATE SYSTEMS

Assume that f is a scalar function and $\boldsymbol{A}$ is a vector function, each having continuous first and second derivatives everywhere within a volume V and on its boundary surface S.

A.4.1 Rectangular Coordinates

The general coordinates are denoted by x, y, and z and the unit vectors by $\boldsymbol{x}$, $\boldsymbol{y}$, and $\boldsymbol{z}$.

$$
\begin{aligned}
\nabla &= \frac{\partial}{\partial x}\boldsymbol{x} + \frac{\partial}{\partial y}\boldsymbol{y} + \frac{\partial}{\partial z}\boldsymbol{z} && \text{(A.16)}\\
\nabla f &= \frac{\partial f}{\partial x}\boldsymbol{x} + \frac{\partial f}{\partial y}\boldsymbol{y} + \frac{\partial f}{\partial z}\boldsymbol{z} && \text{(A.17)}\\
\nabla \cdot \boldsymbol{A} &= \frac{\partial A_x}{\partial x} + \frac{\partial A_y}{\partial y} + \frac{\partial A_z}{\partial z} && \text{(A.18)}
\end{aligned}
$$

$$\nabla \times \boldsymbol{A} = \left(\frac{\partial A_z}{\partial y} - \frac{\partial A_y}{\partial z}\right)\boldsymbol{x} + \left(\frac{\partial A_x}{\partial z} - \frac{\partial A_z}{\partial x}\right)\boldsymbol{y} + \left(\frac{\partial A_y}{\partial x} - \frac{\partial A_x}{\partial y}\right)\boldsymbol{z} \quad \text{(A.19)}$$

$$\nabla^2 f = \frac{\partial^2 f}{\partial x^2} + \frac{\partial^2 f}{\partial y^2} + \frac{\partial^2 f}{\partial z^2} \quad \text{(A.20)}$$

A.4.2 Cylindrical Coordinates

The general coordinates are denoted by ρ, φ, and z and the unit vectors by $\boldsymbol{\rho}$, $\boldsymbol{\varphi}$, and $\boldsymbol{z}$ respectively.

$$\nabla f = \frac{\partial f}{\partial \rho}\boldsymbol{\rho} + \frac{1}{\rho}\frac{\partial f}{\partial \varphi}\boldsymbol{\varphi} + \frac{\partial f}{\partial z}\boldsymbol{z} \quad \text{(A.21)}$$

$$\nabla \cdot \boldsymbol{A} = \frac{1}{\rho}\frac{\partial}{\partial \rho}(\rho A_\rho) + \frac{1}{\rho}\frac{\partial A_\varphi}{\partial \varphi} + \frac{\partial A_z}{\partial z} \quad \text{(A.22)}$$

$$\nabla \times \boldsymbol{A} = \left(\frac{1}{\rho}\frac{\partial A_z}{\partial \varphi} - \frac{\partial A_\varphi}{\partial z}\right)\boldsymbol{\rho} + \left(\frac{\partial A_\rho}{\partial z} - \frac{\partial A_z}{\partial \rho}\right)\boldsymbol{\varphi} + \left[\frac{1}{\rho}\frac{\partial}{\partial \rho}(\rho A_\varphi) - \frac{1}{\rho}\frac{\partial A_\rho}{\partial \varphi}\right]\boldsymbol{z} \quad \text{(A.23)}$$

$$\nabla^2 f = \frac{1}{\rho}\frac{\partial}{\partial \rho}\left(\rho\frac{\partial f}{\partial \rho}\right) + \frac{1}{\rho^2}\frac{\partial^2 f}{\partial \varphi^2} + \frac{\partial^2 f}{\partial z^2} \quad \text{(A.24)}$$

A.4.3 Spherical Coordinates

The general coordinates are denoted by r, θ, and ϕ and the unit vectors by $\boldsymbol{r}$, $\boldsymbol{\theta}$, and $\boldsymbol{\phi}$.

$$\nabla f = \frac{\partial f}{\partial r}\boldsymbol{r} + \frac{1}{r}\frac{\partial f}{\partial \theta}\boldsymbol{\theta} + \frac{1}{r\sin(\theta)}\frac{\partial f}{\partial \varphi}\boldsymbol{\phi} \quad \text{(A.25)}$$

$$\nabla \cdot \boldsymbol{A} = \frac{1}{r^2}\frac{\partial}{\partial r}(r^2 A_r) + \frac{1}{r\sin(\theta)}\frac{\partial}{\partial \theta}(A_\theta \sin\theta) + \frac{1}{r\sin(\theta)}\frac{\partial A_\phi}{\partial \phi} \quad \text{(A.26)}$$

$$\nabla \times \boldsymbol{A} = \frac{1}{r\sin(\theta)}\left[\frac{\partial}{\partial \theta}(A_\phi \sin\theta) - \frac{\partial A_\theta}{\partial \phi}\right]\boldsymbol{r} + \frac{1}{r}\left[\frac{1}{\sin(\theta)}\frac{\partial A_r}{\partial \phi} - \frac{\partial}{\partial r}(rA_\phi)\right]\boldsymbol{\theta} + \frac{1}{r}\left[\frac{\partial}{\partial r}(rA_\theta) - \frac{\partial A_r}{\partial \theta}\right]\boldsymbol{\phi} \quad \text{(A.27)}$$

$$\nabla^2 f = \frac{1}{r^2}\frac{\partial}{\partial r}\left(r^2\frac{\partial f}{\partial r}\right) + \frac{1}{r^2\sin(\theta)}\frac{\partial}{\partial \theta}\left(\sin\theta\frac{\partial f}{\partial \theta}\right) + \frac{1}{r^2\sin^2(\theta)}\frac{\partial^2 f}{\partial \phi^2} \quad \text{(A.28)}$$

Appendix B

Some Useful Integrals

The following integrals may be useful when applying the vector Kirchhoff-Huygens formulas to circular apertures:

$$\int_0^{2\pi} \exp(jx\cos\phi)d\phi = 2\pi J_0(x) \tag{B.1}$$

$$\int_0^{2\pi} \cos(n\phi)\exp(jx\cos\phi)d\phi = j^n 2\pi J_n(x) \tag{B.2}$$

$$\int_0^{2\pi} \cos^2(\phi)\exp(jx\cos\phi)d\phi = \pi\,[J_0(x) - J_2(x)] \tag{B.3}$$

$$\int_0^{2\pi} \sin^2(\phi)\exp(jx\cos\phi)d\phi = \pi\,[J_0(x) + J_2(x)] \tag{B.4}$$

$$\int_0^{2\pi} \sin(n\phi)\exp(jx\cos\phi)d\phi = 0 \tag{B.5}$$

In the above x and ϕ are real continuous variables, n is an integer variable, and $J_n(x)$ is the Bessel function of first kind, order n, and argument x.

Appendix C

Kirchhoff-Huygens Formulas

C.1 KIRCHHOFF-HUYGENS FORMULAS IN BOUNDED MEDIA

Maxwell's equations in the symmetric form will be used here to derive the wave equations in the presence of field sources. Assuming a linear, isotropic, lossless medium and a time dependence of the form $\exp(j\omega t)$ and applying the operator curl to (2.50) and (2.51) we get

$$\nabla \times (\nabla \times \boldsymbol{E}) - k^2 \boldsymbol{E} = -j\omega\mu \boldsymbol{J}_e - \nabla \times \boldsymbol{J}_m \quad \text{(C.1)}$$

$$\nabla \times (\nabla \times \boldsymbol{H}) - k^2 \boldsymbol{H} = -j\omega\varepsilon \boldsymbol{J}_m + \nabla \times \boldsymbol{J}_e \quad \text{(C.2)}$$

where, as usual, $k^2 = \omega^2 \varepsilon \mu$.

These equations can be integrated using Stokes's theorem. The proof given here follows closely Stratton and Chu ([1] as referred in [2]).

Let $\boldsymbol{X}$ and $\boldsymbol{G}$ be the two vector functions of point each continuous and having continuous first and second derivatives everywhere within a volume V and on its boundary surface S (see Figure C.1) where

$$S = \sum_{i=1}^{n} S_i \quad \text{(C.3)}$$

Let $\boldsymbol{n}$ be the unit vector normal at each point on the surface S, pointing toward V.

Considering the vector expression

$$\boldsymbol{X} \times (\nabla \times \boldsymbol{G}) - \boldsymbol{G} \times (\nabla \times \boldsymbol{X}) \quad \text{(C.4)}$$

Stokes's theorem states

$$\int_V \nabla \cdot [\boldsymbol{X} \times (\nabla \times \boldsymbol{G}) - \boldsymbol{G} \times (\nabla \times \boldsymbol{X})] dV = -\int_S [\boldsymbol{X} \times (\nabla \times \boldsymbol{G}) - \boldsymbol{G} \times (\nabla \times \boldsymbol{X})] \cdot \boldsymbol{n} dS \quad \text{(C.5)}$$

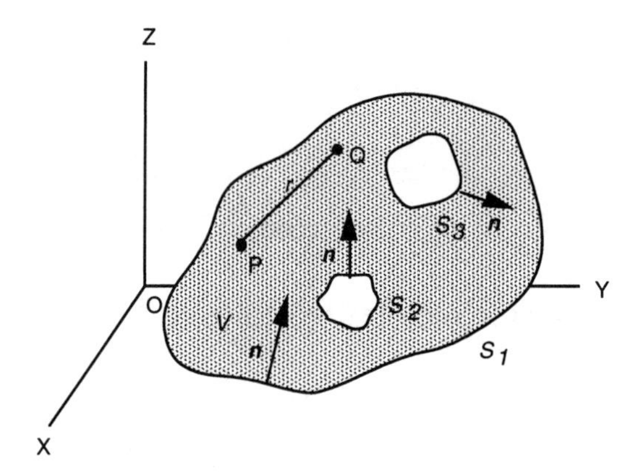

Figure C.1: Geometry for the integration of the wave equations using Stokes's theorem.

Using vector identity (A.6), the left-hand side of (C.5) can be modified to

$$\int_V \{\boldsymbol{X} \cdot [\nabla \times (\nabla \times \boldsymbol{G})] - \boldsymbol{G} \cdot [\nabla \times (\nabla \times \boldsymbol{X})]\}\, dV$$
$$= -\int_S [\boldsymbol{X} \times (\nabla \times \boldsymbol{G}) - \boldsymbol{G} \times (\nabla \times \boldsymbol{X})] \cdot \boldsymbol{n}\, dS \qquad \text{(C.6)}$$

It will be shown now that the electromagnetic fields at a point P, inside volume V can be expressed in terms of the field sources (electric and magnetic charges and current densities) existing inside volume V and the fields on the boundary surfaces.

Let r be the distance between point P and the general point Q inside V. Taking O to be the origin of the coordinate system we have

$$r = |\bar{OQ} - \bar{OP}| \qquad \text{(C.7)}$$

Choosing the function $\boldsymbol{G}$ to be

$$\boldsymbol{G} = \frac{\exp(-jkr)}{r}\boldsymbol{c} \qquad \text{(C.8)}$$
$$= \psi \boldsymbol{c} \qquad \text{(C.9)}$$

where $\boldsymbol{c}$ is an arbitrary but constant vector, then $\boldsymbol{G}$ will obey the continuity conditions everywhere within V except at P. Accordingly, we will exclude from V a volume V_0, enclosed by a sphere with radius r_0 centered on P, the surface S_0 of this sphere being from now on included in the boundary surfaces.

If $\boldsymbol{X}$ is identified with the electrical field $\boldsymbol{E}$, (C.6) becomes

$$\int_{V_1} \{\boldsymbol{E} \cdot [\nabla \times (\nabla \times \psi \boldsymbol{c})] - \psi \boldsymbol{c} \cdot [\nabla \times (\nabla \times \boldsymbol{E})]\}\, dV = \int_{S_1} [\boldsymbol{E} \times (\nabla \times \psi \boldsymbol{c}) - \psi \boldsymbol{c} \times (\nabla \times \boldsymbol{E})] \cdot \boldsymbol{n}\, dS \tag{C.10}$$

where

$$V_1 = V - V_0 \tag{C.11}$$

$$S_1 = S + S_0 \tag{C.12}$$

Using (A.11) and (A.4) we can write

$$\nabla \times [\nabla \times (\psi \boldsymbol{c})] = \nabla[\nabla(\psi) \cdot \boldsymbol{c}] + k^2 \psi \boldsymbol{c} \tag{C.13}$$

Combining (C.1) and (C.13) the integrand on the left-hand side of (C.10) becomes

$$\boldsymbol{E} \cdot [\nabla \times (\nabla \times \psi \boldsymbol{c})] - \psi \boldsymbol{c} \cdot [\nabla \times (\nabla \times \boldsymbol{E})] = \boldsymbol{E} \cdot \nabla(\boldsymbol{c} \cdot \nabla \psi) + \boldsymbol{c} \cdot (\psi \nabla \times \boldsymbol{J}_m + j\omega\mu\psi \boldsymbol{J}_e) \tag{C.14}$$

Taking into account (A.4) and (A.5) we have

$$\nabla \times (\psi \boldsymbol{J}_m) = \psi \nabla \times \boldsymbol{J}_m + \nabla(\psi) \times \boldsymbol{J}_m \tag{C.15}$$

$$\nabla \cdot [(\boldsymbol{c} \cdot \nabla \psi)\boldsymbol{E}] = (\boldsymbol{c} \cdot \nabla \psi) \nabla \cdot \boldsymbol{E} + \boldsymbol{E} \cdot \nabla(\boldsymbol{c} \cdot \nabla \psi) \tag{C.16}$$

Introducing (2.52), (C.14), (C.15), and (C.16), equation (C.10) can be rearranged to yield

$$\int_{V_1} \boldsymbol{c} \cdot \left(j\omega\mu\psi \boldsymbol{J}_e + \boldsymbol{J}_m \times \nabla \psi - \frac{\rho}{\varepsilon} \nabla \psi \right) dV - \int_{S1} (\boldsymbol{c} \cdot \nabla \psi) \boldsymbol{E} \cdot \boldsymbol{n}\, dS - \int_{S1} \boldsymbol{c} \cdot (\psi \boldsymbol{J}_m) \times \boldsymbol{n}\, dS = \int_{S_1} [\boldsymbol{E} \times \nabla \times (\psi \boldsymbol{c}) - \psi \boldsymbol{c} \times \nabla \times \boldsymbol{E}] \cdot \boldsymbol{n}\, dS \tag{C.17}$$

Equation (C.16) can be made independent from $\boldsymbol{c}$. Using (A.5) and (2.50)

$$\boldsymbol{E} \times [\nabla \times (\psi \boldsymbol{c})] \cdot \boldsymbol{n} = \boldsymbol{E} \times (\nabla \psi \times \boldsymbol{c}) \cdot \boldsymbol{n} + [(\boldsymbol{n} \times \boldsymbol{E}) \times \nabla \psi] \cdot \boldsymbol{c} \tag{C.18}$$

$$[\psi \boldsymbol{c} \times (\nabla \times \boldsymbol{E}] \cdot \boldsymbol{n} = \boldsymbol{c} \times (-j\omega\mu\psi \boldsymbol{H} - \psi \boldsymbol{J}_m) \cdot \boldsymbol{n} = j\omega\mu\psi \boldsymbol{c} \cdot (\boldsymbol{n} \times \boldsymbol{H}) + \psi \boldsymbol{c} \cdot (\boldsymbol{n} \times \boldsymbol{J}_m) \tag{C.19}$$

and finally (C.17) gives

$$\mathbf{c}\cdot\left[\int_{V_1}(j\omega\mu\psi \boldsymbol{J}_e + \boldsymbol{J}_m \times \nabla\psi - \frac{\rho}{\varepsilon}\nabla\psi)dV\right]$$
$$= \mathbf{c}\cdot\left\{\int_{S1}[(\boldsymbol{E}\cdot\boldsymbol{n})\nabla\psi + (\boldsymbol{n}\times\boldsymbol{E})\times\nabla\psi - j\omega\mu\psi(\boldsymbol{n}\times\boldsymbol{H})]\,dS\right\} \tag{C.20}$$

Since (C.20) must hold for every vector $\mathbf{c}$, the integrals themselves must be equal. Rearranging (C.20)

$$\int_{S_0}[(\boldsymbol{E}\cdot\boldsymbol{n})\nabla\psi + (\boldsymbol{n}\times\boldsymbol{E})\times\nabla\psi - j\omega\mu\psi(\boldsymbol{n}\times\boldsymbol{H})]\,dS$$
$$= \int_{V_1}(j\omega\mu\psi \boldsymbol{J}_e + \boldsymbol{J}_m \times \nabla\psi - \frac{\rho}{\varepsilon}\nabla\psi)\,dV$$
$$- \int_{S_1}[(\boldsymbol{E}\cdot\boldsymbol{n})\nabla\psi + (\boldsymbol{n}\times\boldsymbol{E})\times\nabla\psi - j\omega\mu\psi(\boldsymbol{n}\times\boldsymbol{H})]\,dS \tag{C.21}$$

We shall now proceed to transform the integral over surface S_0 of the sphere of radius r_0 that encloses the observation point P:

$$\begin{aligned}\nabla\psi &= \nabla\left[\frac{\exp(-jkr)}{r}\right] \\ &= -\left(\frac{1}{r} + jk\right)\frac{\exp(-jkr)}{r}\boldsymbol{r}\end{aligned} \tag{C.22}$$

where $\boldsymbol{r}$ is the unit vector from P to the general point Q. Further noting that $\boldsymbol{r}$ will coincide with the normal to S_0 looking into volume V

$$\int_{S_0}[-j\omega\mu\psi(\boldsymbol{n}\times\boldsymbol{H}) + (\boldsymbol{n}\times\boldsymbol{E})\times\nabla\psi + (\boldsymbol{E}\cdot\boldsymbol{n})\nabla\psi]\,dS$$
$$= \frac{\exp(-jkr_0)}{r_0}\int_{\Omega}\left[-j\omega\mu(\boldsymbol{n}\times\boldsymbol{H}) - (\frac{1}{r_0} + jk)(\boldsymbol{n}\times\boldsymbol{E})\times\boldsymbol{n}\right.$$
$$\left. - (\frac{1}{r_0} + jk)(\boldsymbol{E}\cdot\boldsymbol{n})\boldsymbol{n}\right] r_0^2\,d\Omega \tag{C.23}$$

$d\Omega$ being the element of solid angle subtended by the element of surface dS of the sphere

$$dS = r_0^2 d\Omega \tag{C.24}$$

Rearranging, we get

$$\int_{S_0} [-j\omega\mu\psi(\boldsymbol{n}\times\boldsymbol{H}) + (\boldsymbol{n}\times\boldsymbol{E})\times\nabla\psi + (\boldsymbol{E}\cdot\boldsymbol{n})\nabla\psi]\,dS$$
$$= \exp(-jkr_0)\, r_0 \int_{\Omega} [-j\omega\mu(\boldsymbol{n}\times\boldsymbol{H}) - jk(\boldsymbol{n}\times\boldsymbol{E})\times\boldsymbol{n} - jk(\boldsymbol{E}\cdot\boldsymbol{n})\boldsymbol{n}]\,d\Omega$$
$$- \exp(-jkr_0) \int_{\Omega} [(\boldsymbol{n}\times\boldsymbol{E})\times\boldsymbol{n} + (\boldsymbol{E}\cdot\boldsymbol{n})\boldsymbol{n}]\,d\Omega \tag{C.25}$$

Taking the limit when r_0 tends to zero

$$\int_{S_0} [-j\omega\mu\psi(\boldsymbol{n}\times\boldsymbol{H}) + (\boldsymbol{n}\times\boldsymbol{E})\times\nabla\psi + (\boldsymbol{E}\cdot\boldsymbol{n})\nabla\psi]\,dS = -4\pi\boldsymbol{E}_P \tag{C.26}$$

where $\boldsymbol{E}_P$ designates the vector electric field at a point P. In the limit V_1 tends to V and (C.21) and (C.26) yield

$$\boldsymbol{E}_P = -\frac{1}{4\pi}\int_V \left(j\omega\mu\psi\boldsymbol{J}_e + \boldsymbol{J}_m\times\nabla\psi - \frac{\rho}{\varepsilon}\nabla\psi\right)dV$$
$$+ \frac{1}{4\pi}\int_S [-j\omega\mu\psi(\boldsymbol{n}\times\boldsymbol{H}) + (\boldsymbol{n}\times\boldsymbol{E})\times\nabla\psi + (\boldsymbol{n}\cdot\boldsymbol{E})\nabla\psi]\,dS \tag{C.27}$$

The symmetric form of Maxwell's equations enables us to write immediately the expression for the magnetic field $\boldsymbol{H}_P$ at point P as

$$\boldsymbol{H}_P = -\frac{1}{4\pi}\int_V \left(j\omega\varepsilon\psi\boldsymbol{J}_m - \boldsymbol{J}_e\times\nabla\psi - \frac{\rho_m}{\mu}\nabla\psi\right)dV$$
$$+ \frac{1}{4\pi}\int_S [j\omega\varepsilon\psi(\boldsymbol{n}\times\boldsymbol{E}) + (\boldsymbol{n}\times\boldsymbol{H})\times\nabla\psi + (\boldsymbol{n}\cdot\boldsymbol{H})\nabla\psi]\,dS \tag{C.28}$$

Equations (C.27) and (C.28) are known as the Kirchhoff-Huygens integration formulas. As we have just demonstrated they are exact solutions of Maxwell's equations.

C.1.1 Kirchhoff-Huygens Formulas in Unbounded Media

If volume V is unbounded, but the field sources are all contained in a finite volume, it is important to examine more closely the behavior of the surface integral.

Assuming that the outer boundary is a sphere of radius R, the surface integral in (C.27) becomes

$$\int_{S(R)} [-j\omega\mu\psi(\boldsymbol{n}\times\boldsymbol{H}) + (\boldsymbol{n}\times\boldsymbol{E})\times\nabla\psi + (\boldsymbol{n}\cdot\boldsymbol{E})\nabla\psi]\,dS \tag{C.29}$$

Noting that

$$\begin{aligned}\nabla\psi &= \nabla\left[\frac{\exp(-jkr)}{r}\right] \\ &= -\left(jk+\frac{1}{r}\right)\frac{\exp(-jkr)}{r}\boldsymbol{r}\end{aligned} \tag{C.30}$$

where $\boldsymbol{r}$ is the unit vector normal to the surface of a sphere centered at the origin of the coordinates and pointing outward.

Over the surface $S(R)$ it is

$$\boldsymbol{r} = -\boldsymbol{n} \tag{C.31}$$

Introducing (C.30) and (C.31) in (C.29), we have

$$\int_{S(R)}\left\{j\omega\mu(\boldsymbol{r}\times\boldsymbol{H}) + (jk+\frac{1}{R})\left[(\boldsymbol{r}\times\boldsymbol{E})\times\boldsymbol{r} + (\boldsymbol{r}\cdot\boldsymbol{E})\boldsymbol{r}\right]\right\}\frac{\exp(-jkR)}{R}dS \tag{C.32}$$

which, expanding the double vector product and using the propagation constant definition, becomes

$$\int_{S(R)}\left[j\omega\mu\left(\boldsymbol{r}\times\boldsymbol{H} + \sqrt{\frac{\varepsilon}{\mu}}\boldsymbol{E}\right) + \frac{\boldsymbol{E}}{R}\right]\frac{\exp(-jkR)}{R}dS \tag{C.33}$$

In the limit when R tends to infinity, the surface integral will vanish if the following conditions, also called radiation conditions, are met:

$$\lim_{R\to\infty} R\,|\boldsymbol{E}| \quad \text{is finite} \tag{C.34}$$

$$\lim_{R\to\infty} R\left|(\boldsymbol{r}\times\boldsymbol{H}) + \sqrt{\frac{\varepsilon}{\mu}}\boldsymbol{E}\right| = 0 \tag{C.35}$$

For (C.28), the following conditions should apply:

$$\lim_{R\to\infty} R\,|\boldsymbol{H}| \quad \text{is finite} \tag{C.36}$$

$$\lim_{R\to\infty} R\left|\sqrt{\frac{\varepsilon}{\mu}}(\boldsymbol{r}\times\boldsymbol{E}) - \boldsymbol{H}\right| = 0 \tag{C.37}$$

By an adequate choice of the boundary surfaces it is possible to exclude the field sources from the volume V. If, in addition, this volume is outwardly limited by a sphere of infinite radius, and the fields obey the radiation conditions, we have

$$\boldsymbol{E}_P = \frac{1}{4\pi}\int_S [-j\omega\mu\psi(\boldsymbol{n}\times\boldsymbol{H}) + (\boldsymbol{n}\times\boldsymbol{E})\times\nabla\psi + (\boldsymbol{n}\cdot\boldsymbol{E})\nabla\psi]\,dS \tag{C.38}$$

$$\boldsymbol{H}_P = \frac{1}{4\pi}\int_S [j\omega\varepsilon\psi(\boldsymbol{n}\times\boldsymbol{E}) + (\boldsymbol{n}\times\boldsymbol{H})\times\nabla\psi + (\boldsymbol{n}\cdot\boldsymbol{H})\nabla\psi]\,dS \tag{C.39}$$

S designates the closed surfaces that totally enclose the field sources.

C.2 APERTURE ANTENNAS IN AN UNBOUNDED MEDIUM

In many cases of practical importance, however, it will be impossible (or extremely difficult) to specify the electric and magnetic field over a closed surface. In these cases it will be assumed that the contribution for the integrals (C.38) and (C.39) will come mainly from a portion of surface S, which will be called aperture A. This is in fact equivalent to assuming that everywhere over S, except on aperture A, the fields are zero or that $S - A$ is a perfectly conducting screen with neither currents nor charges.

In this case it is then necessary, in order to ensure the continuity of the electric and magnetic fields over A, to introduce distributions of electric and magnetic charges on the aperture line boundary (Γ_A). The contribution to radiation the fields of such a line of charges can be derived from (C.27) and (C.28):

$$\boldsymbol{E}_P = \frac{1}{4\pi}\int_V \left(\frac{\rho}{\varepsilon}\nabla\psi\right) dV \tag{C.40}$$

$$\boldsymbol{H}_P = \frac{1}{4\pi}\int_V \left(\frac{\rho_m}{\mu}\nabla\psi\right) dV \tag{C.41}$$

Taking χ_e and χ_m, respectively, as the densities of electric and magnetic charge per unit lengths:

$$\chi_e = \lim_{dS\to 0} \rho\, dS \tag{C.42}$$

$$\chi_m = \lim_{dS\to 0} \rho_m\, dS \tag{C.43}$$

and $d\ell$ as the element length along the boundary line:

$$\boldsymbol{E}_P = \frac{1}{4\pi}\oint_{\Gamma_A} \left(\frac{\chi_e}{\varepsilon}\nabla\psi\right) d\ell \tag{C.44}$$

$$\boldsymbol{H}_P = \frac{1}{4\pi}\oint_{\Gamma_A} \left(\frac{\chi_m}{\mu}\nabla\psi\right) d\ell \tag{C.45}$$

Let us now relate χ_e and χ_m to the value of the electric and magnetic fields at the boundary.

Consider, as in Figure C.2, a portion of the aperture A and an elementary rectangular area dS, with sides da and db, respectively, parallel and normal to Γ_A, and the unit vector $\boldsymbol{\tau}$ tangent to Γ_A. Define $\boldsymbol{n}_1$ as a unit vector such that

$$\boldsymbol{n}_1 = \boldsymbol{\tau} \times \boldsymbol{n} \tag{C.46}$$

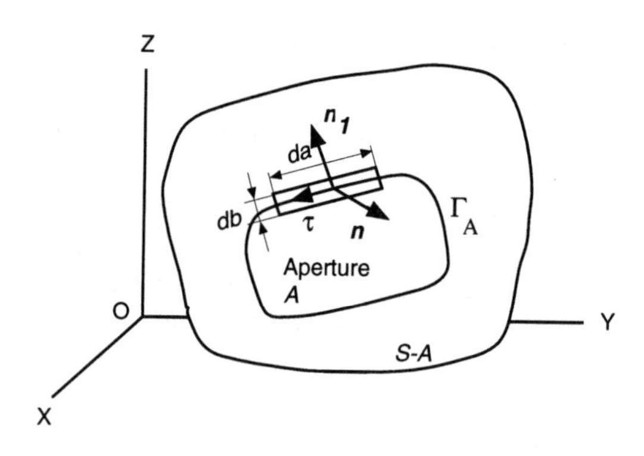

Figure C.2: Boundary conditions: elementary cylinder to apply Gauss's theorem.

The continuity equation for the electric charges on da may be written as

$$(-\boldsymbol{n}_1 \cdot \boldsymbol{J}_e)\, da = -j\omega\chi_e\, da \tag{C.47}$$

where $\boldsymbol{J}_e$ is the surface current density.

For a perfect conductor the equivalence theorem states that

$$\boldsymbol{J}_e = \boldsymbol{n} \times \boldsymbol{H} \tag{C.48}$$

Substituting (C.48) and (C.46) in (C.47) yields

$$\chi_e = -\frac{1}{j\omega}(\boldsymbol{H} \cdot \boldsymbol{\tau}) \tag{C.49}$$

By symmetry we get

$$\chi_m = +\frac{1}{j\omega}(\boldsymbol{E} \cdot \boldsymbol{\tau}) \tag{C.50}$$

The contributions to the total field due to the boundary distributions of electric and magnetic currents are then, from (C.44), (C.45), (C.49), and (C.50),

$$\boldsymbol{E}_P = -\frac{1}{4\pi j\omega\varepsilon}\oint_{\Gamma_A}(\boldsymbol{H}\cdot\boldsymbol{\tau})\nabla\psi\, da \tag{C.51}$$

$$\boldsymbol{H}_P = \frac{1}{4\pi j\omega\mu}\oint_{\Gamma_A}(\boldsymbol{E}\cdot\boldsymbol{\tau})\nabla\psi\, da \tag{C.52}$$

From the applications point of view it is often convenient to transform these line integrals into surface integrals. Taking (C.51)

$$\oint_{\Gamma_A}(\boldsymbol{H}\cdot\boldsymbol{\tau})\nabla\psi\, da = \sum_j \boldsymbol{I}_j \oint_{\Gamma_A}\frac{\partial\psi}{\partial x_j}(\boldsymbol{H}\cdot\boldsymbol{\tau})\, da \tag{C.53}$$

$$= \sum_j \boldsymbol{I}_j \int_S \nabla\times(\frac{\partial\psi}{\partial x_j}\boldsymbol{H})\cdot\boldsymbol{n}\, dS \tag{C.54}$$

where $\boldsymbol{I}_j$, with $j = 1, 2, 3$ are the unit vectors in the direction of the coordinate system axes X, Y, and Z and x_j with $j = 1, 2, 3$ corresponds to x, y and z, respectively.

But from (A.5) it follows that

$$\nabla\times\left(\frac{\partial\psi}{\partial x_j}\boldsymbol{H}\right) = \nabla\left(\frac{\partial\psi}{\partial x_j}\right)\times\boldsymbol{H} + \frac{\partial\psi}{\partial x_j}\nabla\times\boldsymbol{H} \tag{C.55}$$

and then

$$\oint_{\Gamma_A}(\boldsymbol{H}\cdot\boldsymbol{\tau})\nabla\psi dS = \Sigma_j\boldsymbol{I}_j\int_S\left[\left(\nabla\frac{\partial\psi}{\partial x_j}\times\boldsymbol{H}\right)\cdot\boldsymbol{n} + \frac{\partial\psi}{\partial x_j}\nabla\times\boldsymbol{H}\cdot\boldsymbol{n}\right]dS$$

$$= \Sigma_j\boldsymbol{I}_j\int_S\left[-\boldsymbol{n}\times\boldsymbol{H}\cdot\nabla\frac{\partial\psi}{\partial x_j} + j\omega\varepsilon(\boldsymbol{E}\cdot\boldsymbol{n})\frac{\partial\psi}{\partial x_j}\right]dS \tag{C.56}$$

Introducing (C.56) into (C.51) and adding the contributions due to the boundary lines of charge and due to the fields over the aperture A, we get

$$\boldsymbol{E}_P = \frac{1}{4\pi}\int_A\Big[-j\omega\mu\psi(\boldsymbol{n}\times\boldsymbol{H}) + (\boldsymbol{n}\times\boldsymbol{E})\times\nabla\psi + \frac{1}{j\omega\varepsilon}(\Sigma_j\boldsymbol{I}_j(\boldsymbol{n}\times\boldsymbol{H})\cdot\nabla\left(\frac{\partial\psi}{\partial x_j}\right)\Big]dS \tag{C.57}$$

$$\boldsymbol{H}_P = \frac{1}{4\pi}\int_A\Big[+j\omega\varepsilon\psi(\boldsymbol{n}\times\boldsymbol{E}) + (\boldsymbol{n}\times\boldsymbol{H})\times\nabla\psi - \frac{1}{j\omega\mu}(\Sigma_j\boldsymbol{I}_j(\boldsymbol{n}\times\boldsymbol{E})\cdot\nabla\left(\frac{\partial\psi}{\partial x_j}\right)\Big]dS \tag{C.58}$$

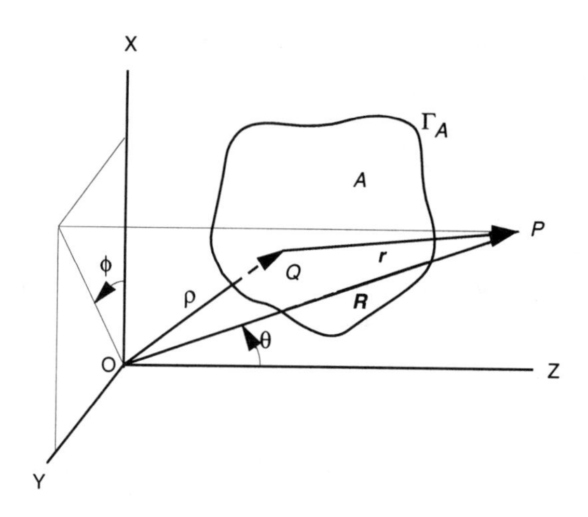

Figure C.3: Geometry for the application of the Kirchhoff-Huygens formulas.

It should be noted, from (C.57) and (C.58), that only the tangential components of the electric and magnetic fields over the aperture are relevant to the radiation fields.

C.3 FAR-FIELD RADIATION OF APERTURE ANTENNAS

Equations (C.57) and (C.58) will now be modified for computational purposes.

Assume a cartesian system of coordinates, a radiating aperture A limited by a contour line Γ_A, a general point Q, also called the integration point, on A and an observation point P which, for future convenience, will be referred by its spherical coordinates R, θ, and ϕ (see Figure C.3).

For large distances from the aperture it may be shown that

$$\lim_{kr\to\infty} (\nabla\psi) = \frac{jk}{r}\exp(-jkr)\,\boldsymbol{r}_1 \tag{C.59}$$

$$\lim_{kr\to\infty} (\nabla\frac{\partial\psi}{\partial x_j}) = -\sum_{i=1,2,3}\frac{k^2}{r} r_{1j}\, r_{1i}\,\exp(-jkr)\,\boldsymbol{I}_i \tag{C.60}$$

$$r \gg \rho \tag{C.61}$$

$$\boldsymbol{R} \quad \text{is parallel to } \boldsymbol{r} \tag{C.62}$$

$$R = r + \boldsymbol{\rho}\cdot\boldsymbol{R}_1 \tag{C.63}$$

where r_j, and $r_1 j$, with $j = 1, 2, 3$ are the components of vectors $\boldsymbol{r}$ and $\boldsymbol{r}_1$, respectively, in the x, y, and z direction. Note that $\boldsymbol{r}_1$ is the unity vector in the direction of $\boldsymbol{r}$.

Substituting (C.59), (C.60), and (C.63) in (C.57) we have

$$\lim_{kr\to\infty} \boldsymbol{E}_P = \frac{1}{4\pi}\int_A \Bigg\{ -j\omega\mu(\boldsymbol{n}\times\boldsymbol{H}) + jk(\boldsymbol{n}\times\boldsymbol{E})\times\boldsymbol{R}_1 - \frac{k^2}{j\omega\varepsilon}\Bigg[\sum_j(\boldsymbol{n}\times\boldsymbol{H}\cdot\boldsymbol{R}_1)R_{1,j}\boldsymbol{I}_j\Bigg]\Bigg\}\frac{\exp -jkr}{r}\,dS \tag{C.64}$$

Because $r >> \rho$, it can be assumed that the variations of inverse of r will be small and therefore

$$\lim_{kr\to\infty} \boldsymbol{E}_P = \frac{\exp -jkR}{4\pi R}\int_A \Bigg[-j\omega\mu(\boldsymbol{n}\times\boldsymbol{H}) - jk\boldsymbol{R}_1\times(\boldsymbol{n}\times\boldsymbol{E}) - \frac{k^2}{j\omega\varepsilon}(\boldsymbol{n}\times\boldsymbol{H}\cdot\boldsymbol{R}_1)\boldsymbol{R}_1\Bigg]\exp(jk\boldsymbol{\rho}\cdot\boldsymbol{R}_1)\,dS \tag{C.65}$$

Considering that

$$\boldsymbol{n}\times\boldsymbol{H} = [(\boldsymbol{n}\times\boldsymbol{H})\times\boldsymbol{R}_1]\times\boldsymbol{R}_1 + [(\boldsymbol{n}\times\boldsymbol{H})\cdot\boldsymbol{R}_1]\,\boldsymbol{R}_1 \tag{C.66}$$

and

$$Z = \sqrt{\frac{\mu}{\varepsilon}} \tag{C.67}$$

we have

$$\lim_{kr\to\infty} \boldsymbol{E}_P = \frac{-jk\exp(-jkR)}{4\pi R}\boldsymbol{R}_1 \times \int_A [\boldsymbol{n}\times\boldsymbol{E} - Z\boldsymbol{R}_1\times(\boldsymbol{n}\times\boldsymbol{H})]\exp(jk\boldsymbol{\rho}\cdot\boldsymbol{R}_1)\,dS \tag{C.68}$$

In the far-field region it is not necessary to compute the magnetic field separately because, as it can easily be shown,

$$\lim_{kr\to\infty}(\boldsymbol{H}_P) = -\frac{\boldsymbol{E}_P}{Z}\times\boldsymbol{R}_1 \tag{C.69}$$

In vacuum

$$\begin{aligned} Z &= Z_0 \\ &= \sqrt{\mu_0/\varepsilon_0} \end{aligned} \tag{C.70}$$

REFERENCES

[1] Stratton, J. A., and Chu, L. J., (1939), Diffraction Theory of Electromagnetic Waves, *Physical Review*, Vol. 56, pp. 99–107.

[2] Silver, S., (ed.), (1965), *Microwave Antenna Theory and Design*, Dover, New-York.

About the Authors

Carlos Salema is a full professor of telecommunication systems at the Instituto Superior Técnico (IST), Technical University of Lisbon, Portugal, since 1980, and director of the Instituto de Telecomunicações since 1993. He graduated in electrical engineering in 1965 at IST and earned a Ph.D. in electrical engineering from the University of London in 1972. First as Chairman of the Board of Directors of the IST Computer Center (1983–85) and later as Chairman of the Board of Directors of the National Foundation for Scientific Computing (1986–89) he was responsible for major improvements in computer facilities in many Portuguese universities and for the initial deployment of the Portuguese scientific data network (RCCN). From 1989 until 1992 he was the Chairman of the Board of Directors of the Portuguese National Science and Technology Research Council (JNICT). During this period JNICT become the most important research funding agency in Portugal and its budget increased from 30 to 150 million US dollars. He has extensive experience of the European Framework Programme activities, having served both as a national delegate, and as technical and strategic auditor.

Carlos Fernandes is an associate professor of microwave antennas and radiowave propagation at Instituto Superior Técnico (IST), Technical University of Lisbon, Portugal, since 1993, and a researcher at Instituto de Telecomunicações, since 1993. He graduated in electrical engineering in 1980 at IST, and earned an M.Sc. in 1985 and a Ph.D. in 1990, in electrical engineering, from IST. Firstly in the scope of RACE project 2067, MBS–Mobile Broadband System. and later in the scope of ACTS project 204, SAMBA, he was responsible for the design, fabrication and testing of dielectric lens antenna systems to provide nearly constant flux illumination of elongated cells at 60 and 40 GHz, respectively.

Rama Kant Jha is a professor of electronics engineering at the Institute of Technology, Banaras Hindu University, India, since 1979. He graduated in Telecommunications Engineering in Jadavpur University, Calcutta, India, in 1964, received an M.Sc. in Electronics and Telecommunications Engineering, from the same university in 1966 and a Ph.D. in Electronics Engineering from Banaras Hindu University, in 1969. His current interest in research include microwave antennas and propagation.

Index

The Artech House Antenna Library

Helmut E. Schrank, Series Editor

Advanced Technology in Satellite Communication Antennas: Electrical and Mechanical Design, Takashi Kitsuregawa

Analysis Methods for Electromagnetic Wave Problems, Volume 2, Eikichi Yamashita, editor

Analysis of Wire Antennas and Scatterers: Software and User's Manual, A. R. Djordjević M. B. Bazdar, G. M. Bazdar, G. M. Vitosevic, T. K. Sarkar, and R. F. Harrington

Antenna-Based Signal Processing Techniques for Radar Systems, Alfonso Farina

Antenna Engineering Using Physical Optics: Practical CAD Techniques and Software, Leo Diaz and Thomas Milligan

Antenna Design With Fiber Optics, A. Kumar

Broadband Patch Antennas, Jean-François Zürcher and Fred E. Gardiol

CAD for Linear and Planar Antenna Arrays of Various Radiating Elements: Software and User's Manual, Miodrag Mikavica and Aleksandar Nešić

CAD of Aperture-fed Microstrip Transmission Lines and Antennas: Software and User's Manual, Naftali Herscovici

CAD of Microstrip Antennas for Wireless Applications, Robert A. Sainati

The CG-FFT Method: Application of Signal Processing Techniques to Electromagnetics, Manuel F. Cátedra, Rafael P. Torres, José Basterrechea, Emilio Gago

Electromagnetic Waves in Chiral and Bi-Isotropic Media, I.V. Lindell, S.A. Tretyakov, A.H. Sihvola, A. J. Viitanen

Fixed and Mobile Terminal Antennas, A. Kumar

Handbook of Antennas for EMC, Thereza Macnamara

IONOPROP: Ionospheric Propagation Assessment Program, Version 1.1: Software and User's Manual, Hernert V. Hitney

Four-Armed Spiral Antennas, Robert G. Corzine and Joseph A. Mosko

Introduction to Electromagnetic Wave Propagation, Paul Rohan

Mobile Antenna Systems, K. Fujimoto and J. R. James

Modern Methods of Reflector Antenna Analysis and Design, Craig Scott

Moment Methods in Antennas and Scattering, Robert C. Hansen, editor

Monopole Elements on Circular Ground Planes, M. M. Weiner et al.

Near-Field Antenna Measurements, D. Slater

Passive Optical Components for Optical Fiber Transmission, Norio Kashima

Phased Array Antenna Handbook, Robert J. Mailloux

Polariztion in Electromagnetic Systems, Warren Stutzman

Practical Simulation of Radar Antennas and Radomes, Herbert L. Hirsch and Douglas C. Grove

Radiowave Propagation and Antennas for Personal Communications, Kazimierz Siwiak

Small-Aperture Radio Direction-Finding, Herndon Jenkins

Solid Dielectric Horn Antennas, Carlos Salema, Carlos Fernandes, and Rama Kant Jha

Spectral Domain Method in Electromagnetics, Craig Scott

Understanding Electromagnetic Scattering Using the Moment Method: A Practical Approach, Randy Bancroft

Waveguide Components for Antenna Feed Systems: Theory and CAD, J. Uher, J. Bornemann, and Uwe Rosenberg

For further information on these and other Artech House titles, including previously considered out-of-print books now available through our In-Print-Forever™ (IPF™) program, contact:

Artech House
685 Canton Street
Norwood, MA 02062
781-769-9750
Fax: 781-769-6334
Telex: 951-659
e-mail: artech@artech-house.com

Artech House
Portland House - Stag Place
London SW1E 5XA England
+44 (0) 171-973-8077
Fax: +44 (0) 171-630-0166
Telex: 951-659
e-mail: artech-uk@artech-house.com

Find us on the World Wide Web at:
www.artech-house.com